Food Fights, Puzzles, and Hideouts

Games, projects, and activities that mix in MATH

Marlene Kliman, Valerie Martin, Nuria Jaumot-Pascual

TERC
2067 Massachusetts Ave
Cambridge, MA 02140
http://terc.edu

Food Fights, Puzzles, and Hideouts **Team**
Author: Marlene Kliman
Design, illustration, and composition: Valerie Martin
Development editor: Nuria Jaumot-Pascual

Tumblehome Learning
201 Newton Street
Weston, MA 02493
www.tumblehomelearning.com

This book was developed at TERC and is based on work funded in part and supported by The National Science Foundation, under grants DRL-07145537, ESI-0406675, and ESI-9901289. Any opinions, findings, and conclusions or recommendations expressed in this material are those of the authors and do not necessarily reflect the views of The National Science Foundation.

TERC is a not-for-profit education research and development organization dedicated to improving mathematics, science, and technology teaching and learning.

Printed in Taiwan
10 9 8 7 6 5 4 3 2 1

ISBN 978-0-9850008-6-8

Table of Contents

Food Fights, Puzzles, and Hideouts

Games, projects, and activities that mix in math

Do you like playing games? building models? eating? talking with friends? running and jumping? exploring and discovering?

If you answered "yes" to any of these, this book is for you. It contains:

Games Galore

All kinds of games: coin, dice, and board; quiet and active; partner and whole group. Play to win or play so everyone wins!

Projects and Crafts

Build, design, create, and grow with projects and crafts that use everyday materials like paper towel tubes and cardboard boxes.

Play with Your Food

Ideas to investigate and games to play when you're eating, cooking, or party planning.

Good for Groups

Icebreakers, party games, challenges, and contests designed for a crowd.

Anytime, Anywhere

Activities to do and games to play wherever you are: in the car, on the bus, in a waiting room, or at the dinner table.

Any Year Calendars

Things to do on familiar holidays (like July 4), less common holidays (like Backward Day) and any day. Includes a bonus set of ideas for celebrating 100 days.

With all the games, projects, and activities in this book, where's the math? It's everywhere: in the strategies you develop when you play the *Name Game* (p. 3), the shapes in the paper tube structures you create with *Build a Hideout* (p. 16), the portion sizes when you play *Food Fight* (p. 31), the distance you go when you participate in *Animal Olympics* (p. 38), the rating you give today's weather when you *Rate It* (p. 52), and the patterns you create with the April calendar (p. 61).

Whether or not you've liked math in the past, once you've dipped into this book, you'll see that it can be part of the things you love to do. So find a game, project, or activity in this book and get started!

Why did we write Food Fights, Puzzles, and Hideouts?

We believe that creativity, play, and socializing are important ingredients in learning just about anything. This book is designed to put those ingredients into learning math.

We started with activities, games, and projects that children do for fun at after-school programs, at public libraries, at school, and at home. We highlighted the inherent math with things to talk about, and sometimes we added a mathematical twist. To ensure that our materials were engaging and enriching, we piloted them in a wide range of "informal" settings (after-school, home, public library) and "formal" programs (academic support, tutoring, school). Independent research showed that children and adults gained math skills, confidence in their math abilities, understanding of the role of math in everyday life, and positive attitudes toward math. To find out more about this research, see http://mixinginmath.terc.edu/aboutMiM/index.php.

Food Fights, Puzzles, and Hideouts is based on nearly 15 years of development and research funded in part by The National Science Foundation.

Who is this book for?

Everyone! It's for children and adults, mathophobes and mathophiles, and parents, teachers, after-school providers, and childcare providers. The games, projects, and activities are geared toward children in the elementary grades, but older children and adults will also enjoy and find challenge in them. Some are perfect for children alone or in small groups; others work well with a larger group at home, at a party, at an after-school program, at school, or just about anywhere!

What math is in the book?

The activities, games, and projects in this book span the key topics in the elementary grades Common Core State Standards for Mathematics. See pages 71-77 for more detail. Many of the ideas in this book are interdisciplinary, including topics in engineering, science, social studies, and literacy.

Thank you!

The authors of the book, Marlene Kliman, Valerie Martin, and Nuria Jaumot-Pascual, are very grateful to Martha Merson and Lily Ko at TERC for their contributions, to our external evaluators who have provided us with evidence-based insights (Char Associates, Miller-Midzik Research Associates, and Program Evaluation Research Group), to Laura DeSantis for amazing images, and to the many after-school providers, librarians, parents, and children who have collaborated with us over the years. We extend our appreciation to TERC for providing a home for our Mixing in Math projects. Marlene would like to thank her daughters Clara and Chloe for helping her mix in math from the start and for providing a reality check on math at home.

Making the most of Food Fights, Puzzles, and Hideouts

This book contains hundreds of games, activities, and projects, including variations and calendar ideas. You can do them in any order.

Look for the information below in each game, project, and activity.

Level. Each is marked with one or more of "Easy," "Medium," and "Hard." Some people use these levels as a starting point: Kindergarteners and first graders might start with "Easy;" fourth and fifth graders with "Hard." Other people prefer to start with "Easy" for almost any age and then move up as needed, since abilities at different ages vary widely.

Note: Levels that only appear in Variations are in parentheses.

Group size. Some projects and activities can be done alone; others are best done with a group. Games require more than one player.

Materials. Some involve no materials; others rely on common household materials.

Talk About. Offers ideas on things to talk over or think about.

Party Planning

Plan a party or special snack to fit your budget.

Levels: Medium (Hard)

Group size: small enough for everyone to have a say in the list

Materials:

- grocery store circulars or access to online grocery store price listing
- paper and pencil
- calculators

1 How many and how much?

Find out how many people are coming and the total you can spend.

How much can you spend per person?
Is $1 each enough?

Talk About...

2 What would you buy?

Look through grocery store circulars or go online to view price lists.

Make a list of what you'll buy.

How did you make your choices?

How did you stay within budget?

We're spending $20. Our list first came to $21.73. We took out a box of crackers for $1.99.

3 Shop, cook, serve, and eat

Take your list to the store and shop. Enjoy the party!

Variations

Party favors (Hard). Figure in the cost everything you'll need for the party. Decide how much you can spend per person total on food, place settings, take-home bags, and other party favors.

Family dinner (Medium). Decide how much you will spend per person or how much in total for a family dinner. Then, plan the menu, and shop, cook, and eat!

About the Authors

Marlene Kliman, Senior Scientist and Director of the Mixing in Math group at TERC, brings 30 years experience developing research-based resources for children's math learning in and out of school. A Principal Investigator of out-of-school math projects funded by The National Science Foundation, she has collaborated with a wide range of educational organizations including after-school programs, public libraries, and family literacy centers. She formerly taught math to pre-service elementary grades teachers at Lesley University. Marlene completed her undergraduate studies in mathematics at Harvard and her graduate studies in learning and epistemology at MIT.

Valerie Martin, Senior Web and Graphic Designer at TERC, specializes in conveying math and science concepts in a clear and visually appealing manner. She designs web- and print-based curricula, games, and educational resources for a wide range of audiences, including children, parents, preschool teachers, adult education teachers, and museum educators. Valerie holds degrees from SUNY Binghamton in French and German literature, with further studies in graphic design and website design and development.

Nuria Jaumot-Pascual, Senior Research Associate at TERC, has 20 years experience as a bilingual (Spanish/English) preschool and after-school teacher, staff developer, and educational researcher in Spain, Central America, and the United States. Nuria holds degrees in out of school education (Universitat de Barcelona), anthropology (University of Texas, Austin), educational psychology (Universitat Oberta de Catalunya), and organizational behavior (Harvard). She is currently a doctoral student in education at the University of Georgia.

Games Galore

All kinds of games: coin, dice, and board; quiet and active; partner and whole group. Play to win or play so everyone wins!

Contents

Games Galore in other sections

Blockade

Block the other players before they block you!

Levels: Medium (Hard)

Group size: 2-3 per game

Materials:

1 piece of graph paper

pencil for each player

Set up for the game

Make the game board. Box off 12 grid squares across and 16 down.

Take turns. On your turn:

❶ Make a rectangle using 12 squares and mark it on the board.

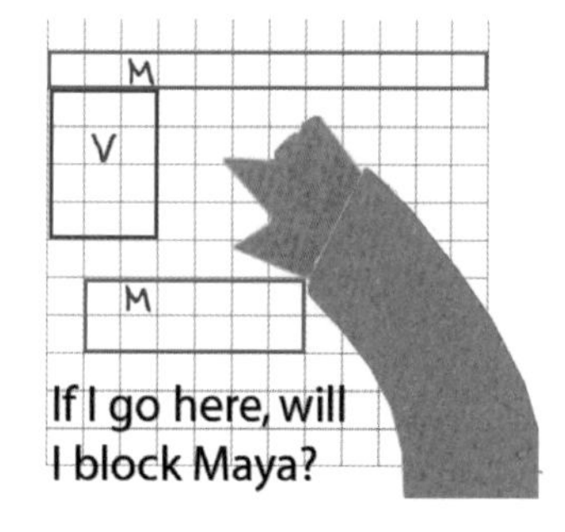

How did you figure out how long and how wide to make your rectangles?

Did you block another player? How?

Talk About...

❷ Keep going until there is no room to take a turn.

The last player to make a rectangle wins.

Variations

Everyone wins (Medium). Work together to fill the game board with as many 12-square rectangles as possible.

Blockade 24 (Medium). Make rectangles with 24 squares.

Six squares (Hard). Take turns drawing shapes with six squares. The squares in the shape must share at least one side. Your shape must be different from all the others. The last person to make a unique shape wins.

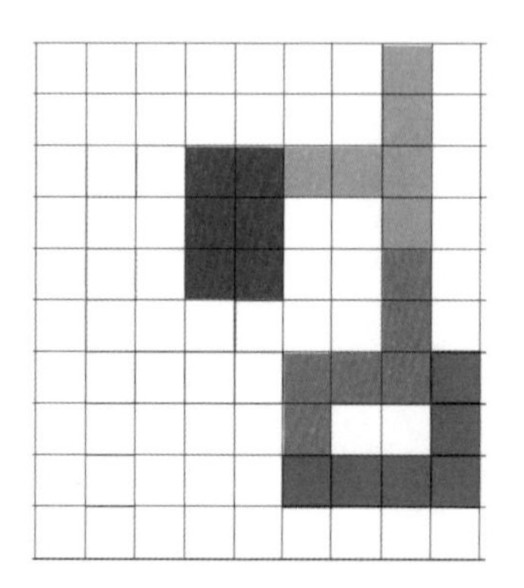

Name Game

Write your name as many times as you can on the game board. Last person to write wins.

Levels: Medium (Hard)

Group size: 2-3 per game

Materials:

1 piece of graph paper

pencil for each player

Set up for the game

Make the game board. Box off 12 squares across and 12 down on the graph paper.

Take turns. On your turn:

1. Write your name in the grid. Names go across or down, with one letter in each square.

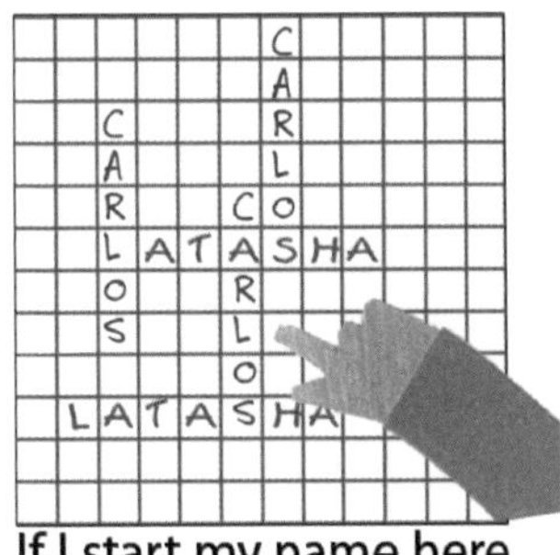

If I start my name here, I'll block Carlos.

How many times does your name fit across the grid?

Did you block another player? How?

2. Keep going until there is no room to take a turn.

The last player to write a name wins.

Variations

Nicknames (Medium). For less challenge, use a nickname or a shorter name.

Different grids (Hard). Play with a 9 × 9 or 15 × 15 game board. Which size makes the game easier? harder?

Everyone wins (Medium, Hard). Work together to try to fill up the game board with your names.

Secret Number

Gather clues to find the secret number.

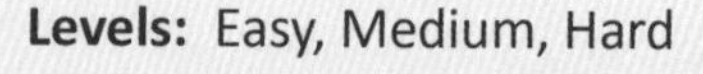

Levels: Easy, Medium, Hard

Group size: 3-5 per game; see Variations for a 2-player game

Materials:

paper and pencil for each player

Set up for the game

Decide who will be the Leader. The Leader secretly picks a number

Easy. Between 1 and 20

Medium. Between 1 and 50

Hard. Between 1 and 100

and announces the range: "I'm thinking of a number between 1 and 50."

Players jot down the numbers in the range.

1	2	3	4	5	6	7	8	9	10
11	12	13	14	15	16	17	18	19	20
21	22	23	24	25	26	27	28	29	30
31	32	33	34	35	36	37	38	39	

Take turns. On your turn:

1 Ask a yes-or-no question to help figure out the secret number. You may not ask if a certain number is the secret one.

The Leader answers the question.

Players cross out the numbers that were ruled out.

2 Keep going until someone finds the secret number.

Talk About...

Can you think of a question that will rule out at least three numbers, whether the answer is "yes" or "no"?

If you ask, "Is the number even?" and the answer is "no," what numbers do you rule out?

The player who identifies the secret number wins.

Variations

Two players (Easy, Medium, Hard). Play two games. Each game, a different person is The Leader. The person who finds the secret number with the fewest questions wins.

Everyone wins (Easy, Medium, Hard). Work together to try to find the secret number with the fewest questions possible.

Roll the Bank

Feeling lucky? Roll the die and try to clear out the bank.

Levels: Medium (Hard)

Group size: 2-4 per game

Materials:

1 die or number cube

masking tape

pen

pennies or small objects such as paper clips or buttons: 10 per player and 10 more for the bank

Before beginning

Tape over each face of the die. Write in: +1, -1, 0, +1/2, -1/2, and Free.

Set up for the game

Each player gets 10 pennies.

Put 10 pennies in the center to start the penny bank.

Take turns. On your turn:

1. Roll the die.
2. If you roll

+1	Take a penny from the bank.
-1	Put a penny in the bank.
0	Do nothing. Your turn is over.
+1/2	Take half* the pennies in the bank.
-1/2	Put half* your pennies in the bank.
Free	You may roll again if you wish.

3. **If you have no more pennies, you are out of the game.**
4. **If you take the last penny from the bank, you win.**

When you rolled "Free," how did you decide whether or not to roll again?

Variation

Pick your half (Hard). If you roll +1/2, take half the pennies in the bank or half of the pennies of any one player.

*For an odd number of pennies, round up to the next even number before finding half. For instance, if there are 7 pennies, find half of 8. For 1 penny, find half of 2.

Heads Up

Toss the coins and find your total. Highest total wins.

Levels: Easy, Medium

Group size: 2 per game

Materials:

Easy. 4-5 pennies and 1-3 other coins

Medium. 6-8 coins of different values

Play five rounds

The player who wins the most rounds wins the game. To play a round follow the steps below.

Take turns. On your turn:

1. Gently toss the coins.

Find the total value of the coins that land heads up. Don't count the coins that land tails up.

How did you find your total? Did you count? add? Talk About...

2. Compare totals.

The player with the larger total wins the round. If you tie, play an extra round.

Variations

Roll 100 (Medium). Keep taking turns until one player has rolled a total of 100 face-up. The first player to reach 100 wins.

Everyone wins (Easy, Medium). Play until each person wins at least three rounds.

Plan ahead to get the last penny. That way, you'll always win!

Levels: Easy (Medium)

Group size: 2 per game

Materials:

10 pennies, paper clips, or other small objects

Set up for the game

Spread out the pennies.

Take turns. On your turn:

1. Take one, two, or three pennies.

   How did you decide how many to take?

2. Keep going until there are no pennies left.

The player who takes the last penny wins.

Variations

Play in rows (Medium). Play with 12 pennies. Put them in three rows: a row of three, a row of four, and a row of five. On each turn, a player may pick up one, two, or three pennies from a single row.

Opposite goal (Easy, Medium). The player who takes the last penny loses, instead of wins. How do you decide how many pennies to take?

Twenty Pennies

Be the first to get 20 pennies.

Set up for the game

Put a penny out to start the penny bank.

Take turns

You get one free pass per game. When you pass, you skip your turn.

On your turn:

1. Roll the die.
2. Read the number. Put that many pennies in the bank. If the total in the bank is:
   - less than 20—the next player takes a turn.
   - 20—the game is over. You win!
   - more than 20—the game is over. The other player wins.

Level: Easy

Group size: 2 per game

Materials:

1 die or number cube

20 pennies or small objects such as paper clips or buttons

How did you decide when to use your free pass?

How did you organize the pennies so they're easy to count?

Talk About...

Variation

Up to three (Easy). Use maskng tape to cover the 4, 5, and 6 on the dice. Write in 1, 2, and 3, so each die has two 1s, two 2s, and two 3s.

Same as Seven

Use your dice roll to make a number as close as possible to 7. Closest wins.

Levels: Medium, Hard

Group size: 2-3 per game

Materials:

3 dice or number cubes

pencil and paper for each player

❶ Roll

One player rolls the dice.

❷ Write

Medium. Each player uses the numbers rolled and + and - to write an expression as close as possible to 7.

Hard. Players may also use × and ÷.

❸ Compare

Your score is the difference between the number you make and 7. Lowest score wins.

Variations

Best in three (Medium, Hard). Play three rounds. Lowest total score wins.

Negative and positive (Hard). Use negative and positive scores. If you make 5, score -2. If you make 9, score +2. Score closest to 0 wins.

Change the dice (Hard). Put tape over some of the numbers on the dice and write in new numbers. To make the game harder, change 1 to 10, 2 to -1, and 3 to 0. Or, play with dice that have 4, 10, or 12 sides.

Land on 100

Be the first to reach 100 on the game board.

Levels: Medium (Easy)

Group size: 2-3 per game

Materials:

3 dice or number cubes

penny or button

Land on 100 Game Board

Set up for the game

Put the penny on number 1 on the game board.

Take turns. On your turn:

1. Roll one, two, or three dice (your choice).

   How did you decide how many dice to roll?

   Talk About...

2. Find the total of your roll.
3. Add your total to the number the penny is on.

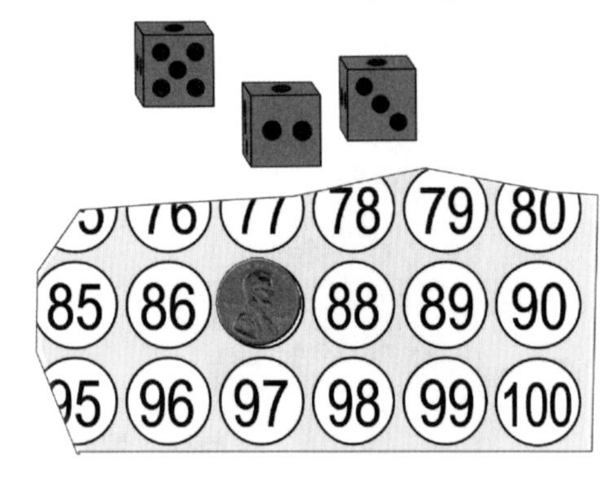

- If the new number is less than 100, move the penny there.
- If the new number is exactly 100, move to 100. The game is over. You won!
- If the new number is more than 100, leave the penny where it is.

Variations

Play with one die (Easy). Roll just one die each turn. Move that number of spaces. The first person to land on 100 wins.

Change the dice (Easy, Medium). Put tape over some of the numbers on the dice and write in new numbers. To make the game easier, change 4, 5, and 6 to 1, 2, and 3, so you have two of each number. To make the game harder, change 1 to 10, 2 to -1, and 3 to 0. Or, play with dice that have 4, 10, or 12 sides.

Land on 100 Game Board

1	2	3	4	5	6	7	8	9	10
11	12	13	14	15	16	17	18	19	20
21	22	23	24	25	26	27	28	29	30
31	32	33	34	35	36	37	38	39	40
41	42	43	44	45	46	47	48	49	50
51	52	53	54	55	56	57	58	59	60
61	62	63	64	65	66	67	68	69	70
71	72	73	74	75	76	77	78	79	80
81	82	83	84	85	86	87	88	89	90
91	92	93	94	95	96	97	98	99	100

More, Less, or Equal

Heads or tails—how will the coins land? Make the most correct predictions to win.

Levels: Easy, Medium

Group size: 2 per game

Materials:

Easy. 5-7 coins equal to about 20 cents in total

Medium. 5-7 coins equal to about $1.00 in total

Play five rounds

To play a round, follow the steps below.

1. **Predict your total.**

You are going to toss the coins. Predict: Will the total of the coins that land heads up be more than, less than, or equal to

Easy. 10 cents?

Medium. 50 cents?

2. **Tell the other player your prediction.**

3. **Gently toss the coins.**

4. **Find the total of the coins that land heads up.**

5. **If you predicted correctly, score a point for the round.**

The player with the most points after five rounds wins the game.

Variations

Play with pennies (Easy). Use six pennies. Predict whether the total of the pennies that land heads up will be more than, less than, or equal to 3 cents.

Everyone wins (Easy, Medium). Play until each person has at least three points.

Narrow It Down

Does it grow on trees? Is it safe to eat? Gather clues to find the secret object.

Levels: Easy, Medium, Hard

Group size: 3-5 per game; see Variations for a 2-player game

Materials:

assortment of objects per group:

Easy. 8-10 objects

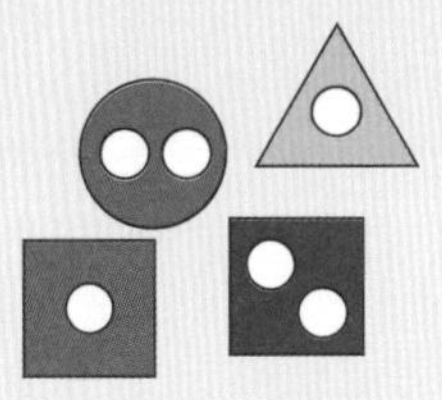

Medium. 10-20 varied everyday objects

Hard. 20-30 objects that vary in subtle ways, such as coins or flags from different countries

Set up for the game

Spread out the objects.

Decide who will be the Leader. The Leader secretly picks an object without removing it from the layout.

Take turns. On your turn:

1 Ask a yes-or-no question to help figure out the secret object.

You may not ask if a certain object is the secret one.

Does it have a diagonal line?

Does it have parallel lines?

Is it symmetrical?

2 The Leader answers the question and removes any objects that were ruled out.

What's a yes-or-no question that could rule out about half of the objects?

How do you decide what to remove if the answer is "no"?

The player who narrows the layout down to the secret object wins.

Variations

Two players (Medium). Play two games. Each game, a different person is The Leader. The person who identifies the secret object with the fewest questions wins.

(continued on next page)

Guess who (Easy, Medium). Play with at least six people. The Leader secretly picks one person in the group. To narrow down to the secret person, players ask questions such as, "Is this person wearing a shirt with a triangle on the front?" or "Is this person's hair more than 12 inches long?"

Name that shape (Easy, Medium, Hard). The Leader secretly picks an object in the room that everyone can see and announces the shape ("I see a rectangle"). Players ask yes-or-no questions to try to identify the secret object.

Projects and Crafts

Build, design, create, and grow with projects and crafts that use everyday materials like paper towel tubes and cardboard boxes.

Contents

Projects and Crafts in other sections

Build a Hideout

Want to get away? Build a hideout you can fit inside.

❶ Build

Use the materials to build a hideout that:

- **Easy.** Fits one person.
- **Medium.** Fits two people.
- **Hard.** Fits one person lying down.

Levels: Easy, Medium, Hard

Group size: 1–3 per hideout

Materials per hideout:

- 40 or more cardboard tubes from paper towel rolls (or, make tubes from tightly rolled newspaper and tape)
- scissors
- tape

❷ Give tours

What shapes did you use to make your hideout?
How did you make sure it is sturdy?

❸ Hideout awards (optional)

Each hideout wins an award for size or shape. Everyone decides on awards together.

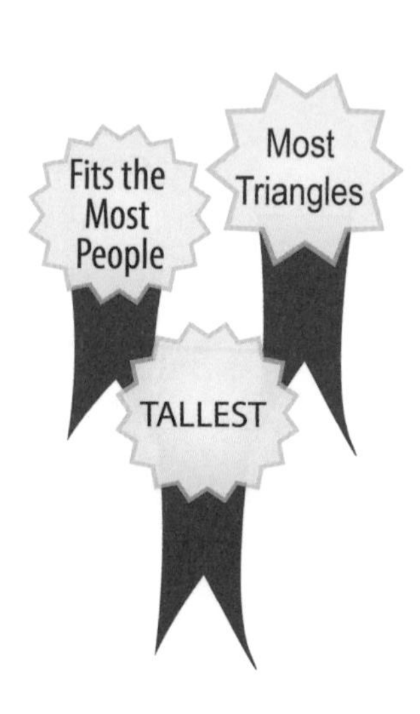

Variations

Hide out in a tent (Easy, Medium, Hard). Make your hideout strong enough so that you can cover it with a sheet or blanket.

Toy hideout (Medium, Hard). Use toothpicks and gumdrops to build a hideout for a toy animal or person. The hideout should have at least 4 square inches of floor space. For extra challenge, build a hideout that encloses at least 12 cubic inches of space.

Potato Bridge

Pile on the potatoes! Build a cardboard bridge that holds as many potatoes as possible.

Levels: Easy, Medium

Group size: 1-2 per bridge

Materials per bridge:

- cardboard from two cereal boxes
- scissors
- tape
- 5 lb. bag of potatoes (can be shared for more than one bridge)
- ruler (Medium)

❶ Build, test, and revise

Use the materials to build a bridge that:

Easy. Holds as many potatoes as possible.

Medium. Holds as many potatoes as possible and is at least 8 inches high.

How can you keep your bridge from collapsing when you add an extra potato?

❷ Make it stronger

Try to build a bridge that holds even more potatoes. For extra challenge, use only one cereal box.

❸ Bridge awards (optional)

Each bridge wins an award for size or strength. Everyone decides on awards together.

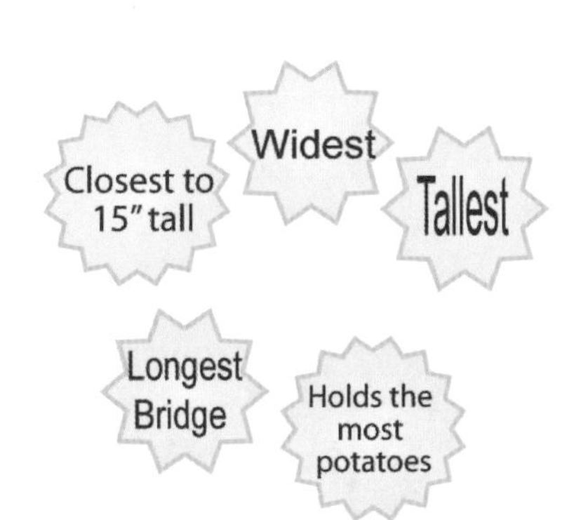

Variations

Use tubes (Easy). In addition to the cardboard from cereal boxes, use up to two paper towel tubes to make your bridge.

Gumdrop bridge (Easy, Medium). Build the strongest bridge you can with 50 gumdrops and 200 toothpicks. For extra challenge, make the bridge at least 10 inches long.

Ride on a Slide

Build a cardboard slide that takes three seconds to roll down.

❶ Build, test, and revise

Use the materials to build a slide that takes your "slide rider" three seconds to roll down.

Your "slide rider" should stay on the slide the whole time. No falling off!

How can you adjust the slide so the rider slows down?

How can you change the slide so the rider speeds up?

Talk About...

Levels: Medium (Hard)

Group size: 1–3 per slide

Materials per slide:

2 boxes (cereal or shoe box size), taken apart so they remain in one piece

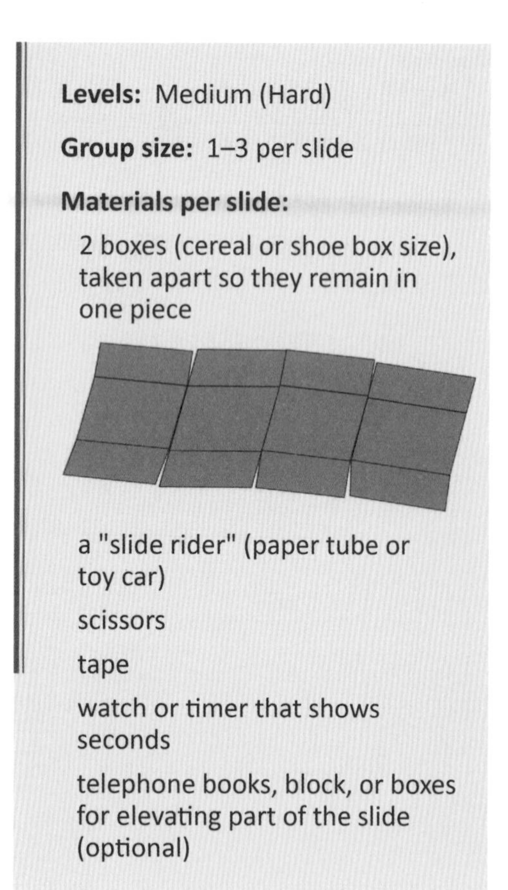

a "slide rider" (paper tube or toy car)

scissors

tape

watch or timer that shows seconds

telephone books, block, or boxes for elevating part of the slide (optional)

❷ Show off your slide

Demonstrate the slide for someone else.

Would your slide give a safe ride? Why or why not?

Talk About...

❸ Slide awards (optional)

Each slide wins an award for speed or slope. Everyone decides on awards together.

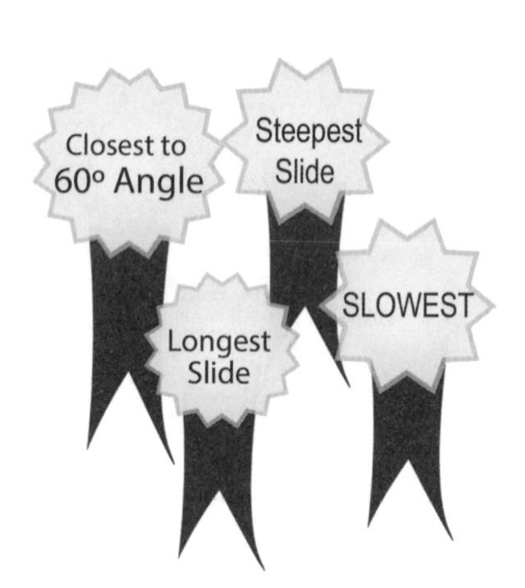

Variations

Use tubes (Medium). In addition to the cardboard, use up to two paper towel tubes to make your slide.

Make it steep and scary (Hard). Start your slide at least 18 inches from the ground.

Slide uphill (Hard). Build your slide so that the "slide rider" goes uphill for part of the ride.

Tell Me a Story

Make your own counting book and share it with someone else.

Levels: Easy, Medium, Hard

Group size: at least one person to make a book and one to hear the story

Materials per book:

several sheets of paper (optional: fold and staple to make a book)

markers, stickers, and other art supplies

1 Decide how you'll count and how high you'll go

Easy. Count up by 1, 2 or 1¢. Start at 0. Or, count back and start at 10 (or 10¢).

Medium. Count up by 4, 5, 10, or 10¢. Start at 0. Or, count back from 50 or 100.

Hard. Count up or back by 25¢, 1/2, 3/2, 7, or 11. Start anywhere.

2 Plan your story

What happens in the story? What increases (or decreases) on each page?

3 Put it on paper

Write your story, or plan it out so you can tell it as you show the pages. Each page of the book should include a number and a picture to match.

4 Read or tell your story

What number will be on the next page? What about on the page after that? Why do you think so?

Variation

Square stories (Hard). Write a counting book based on square numbers: 1 (1 × 1); 4 (2 × 2); 9 (3 × 3).

Estimation Station

Make a mystery cup. Fill it up, and challenge others to find out how many are inside.

Levels: Easy, Medium

Group size: 2 or more (to make and trade cups)

Materials (per cup):

- clear plastic cup
- rubber band
- sheet of plastic wrap
- objects for filling cups
  - **Easy**. large objects (e.g., big pieces of pasta) so 10-20 fill the cup
  - **Medium**. smaller objects (e.g., beads) so 20-50 fill the cup

❶ Count and fill

Count out objects to fill a cup. Keep the total a secret!

Talk About...

How can you group the objects so they're easy to count?

If you count them by 2s and then count them by 5s, will you get the same answer?

❷ Cover the cup with plastic wrap

Secure the wrap with a rubber band.

❸ Trade cups

Estimate how many are inside. Count to check.

Variation

Mystery snack (Easy, Medium). Make a mystery cup filled with dried fruit or another food. Everyone estimates and counts before eating.

Estimation Station Challenge

How do you estimate? Use what you know about one cup to find how many in the other.

Level: Hard

Group size: 2 or more (to make and trade cups)

Materials (per two cups):

- two clear plastic cups the same size
- two rubber bands
- two sheets of plastic wrap
- objects that come in two sizes (pompoms, pasta shells): larger so that about 20-40 fill the cup, and smaller so that at least 100 fill the cup

❶ Make two mystery cups

Follow steps 1 and 2 from Estimation Station. Put the larger objects in Cup 1 and the smaller in Cup 2.

❷ Trade cups

Say how many are in Cup 1. The other person estimates how many are in Cup 2.

Cup 1 has 30. I estimate 60 in Cup 2 because the pompoms are smaller.

Does Cup 2 have about twice as many as Cup 1? about 10 times as many?

I think Cup 2 has 150 because one big equals about five small ones.

180. I did length times width then I multiplied that by height.

Growing Plants

How fast does your garden grow? Grow plants and track how they grow over time.

Before beginning

Plant seeds or seedlings.

❶ Predict

How tall will your plant be in a week? How tall in a month?

❷ Measure each week

Easy. Cut straws to the height of the plant. Tape the straws to graph paper.

Medium. Measure with a ruler and mark the height on graph paper.

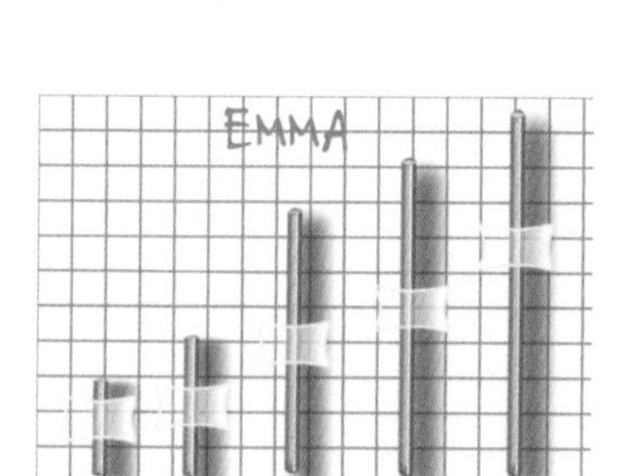

❸ What do you notice?

Does your plant grow about the same amount each week? How can you tell?

How does the actual plant growth compare with your predictions?

Levels: Easy, Medium

Group size: 1-2 per plant

Materials per plant:

- seeds or seedlings that grow quickly (e.g., grass, beans, avocado pit)
- potting soil and a pot
- a few straws
- scissors
- tape
- marker
- piece of graph paper
- ruler (Medium)

Variation

Change the conditions (Medium). Put one plant in the sun and one in the shade. Do both grow about the same amount each week?

Giant Size

If an eraser were ten times its size, could you hide behind it? Make one to find out!

Levels: Medium, Hard

Group size: 1-2 per giant object

Materials per giant object:

large sheet of graph or plain paper (or, tape a few sheets together to make a big piece)

flat, rectangular object such as a dollar or eraser

pencil, marker, scissors

ruler or yardstick (Hard)

❶ Could you hide behind a giant one?

Pick a flat, rectangular object such as an eraser or a dollar.

Imagine that is ten times longer and ten times wider.
Then predict:

- Would it cover your hand?
- Would it cover your face?
- Could you hide behind it?

❷ Make a giant one

Medium. Trace your object ten times across and ten times down.

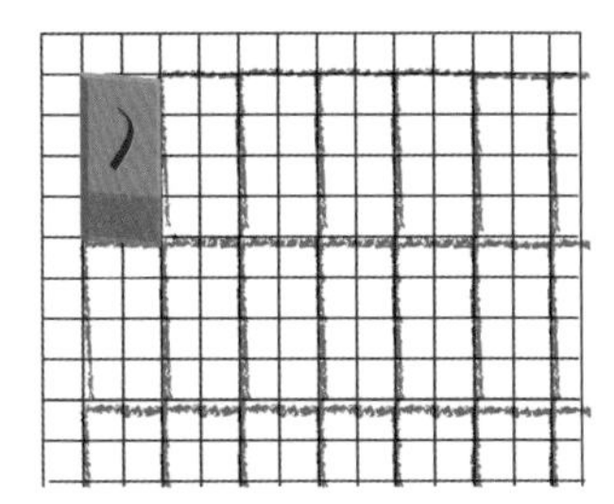

Hard. Measure your object, multiply length and width by ten, and draw the giant one on graph paper.

Cut out and decorate your giant object.

❸ Compare predictions and results

Can you cover your face with it? Hide behind it?

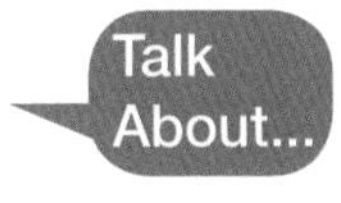

Variation

Could you fit inside it? (Hard). Pick a three-dimensional object such as a cup or juice box. Predict: Could you fit inside it if it were ten times wider, longer, and higher? Make one and see!

Penny Jar

Save your spare change for a special purchase or donation.

Levels: Medium, Hard (Easy)

Group size: small enough so everyone gets a chance to count coins

Materials:

clear jar for coins

paper and pencils

calculator (optional)

Before beginning

Choose something to save up for.

Medium. Choose something that costs up to $10.

Hard. Choose something that costs up to $100.

❶ Predict

How long do you think it will take for us to reach our goal?

How full will the jar be when we reach the goal?

Talk About...

Write down your predictions and today's date.

❷ Collect and count

Put spare change in the jar every day.

Count the change every week and write down the amount.

APRIL

day	Wednesday	Thursday	Friday	Saturday
	1	2 $1.85	3	4
7	8	9 $4.07	10	11
14	15	16	17	18

We collected 407 pennies this week. What's that in dollars?

Compare your predictions with your progress so far.

Keep counting and collecting until you reach the goal.

Variations

Make tens (Easy). Sort coins by type. Then, put them in piles of ten. An older child or adult helps find the total.

How long to fill the jar? (Medium, Hard). Predict how long it will take to fill the jar with spare change. Then, try it and count up how much you have when the jar is full.

Puzzle Me This

Stump your friends with a puzzle you design.

Levels: Hard (Medium)

Group size: 2 or more (to make and trade puzzles)

Materials per puzzle:

- piece of graph paper
- scissors
- pencil
- envelope
- colored pencils (optional)
- thick cardboard or foam for puzzle backing; glue (optional)

❶ Make the puzzle grid

Block off 12 grid squares across and 12 down.

Cut out the 12 × 12 square.

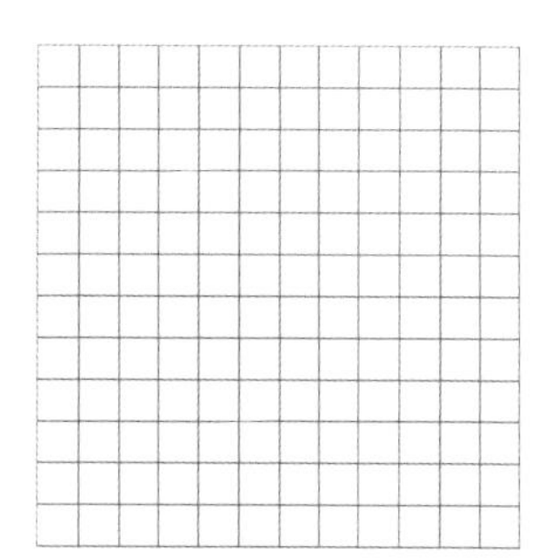

❷ Plan the pieces

Plan how you will divide the 12 × 12 square into six pieces. Each piece should be made from the same number of squares but have a different shape.

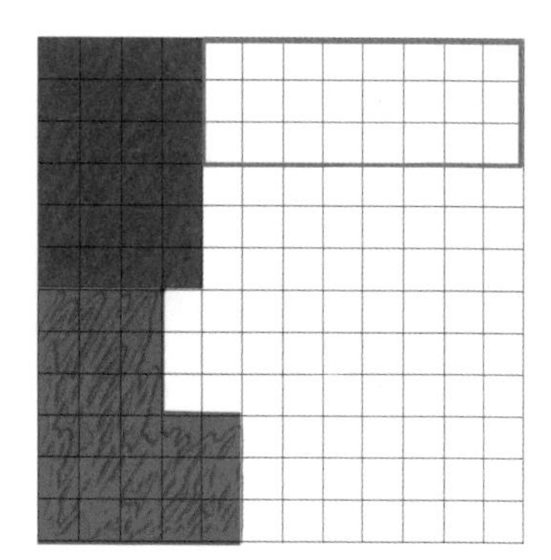

Draw the outline of the pieces on the 12 × 12 square.

Color in the pieces (optional).

Glue the puzzle onto a cardboard or foam backing (optional).

❸ Cut

Put the six pieces in an envelope.

❹ Trade puzzles

Can you find more than one way to assemble the pieces into a 12 × 12 square?

(continued on next page)

Variations

Two shapes (Medium). Divide the 12 × 12 square into six pieces: three of one shape and three of another.

Double trouble (Hard). Make and trade 12 piece puzzles. Divide the six pieces in half to make 12 pieces.

Different size pieces (Hard). Block off 16 squares across and 16 down. The largest piece should use half the squares in the grid, the next largest should use half of that, the next-largest is half of that, and so on.

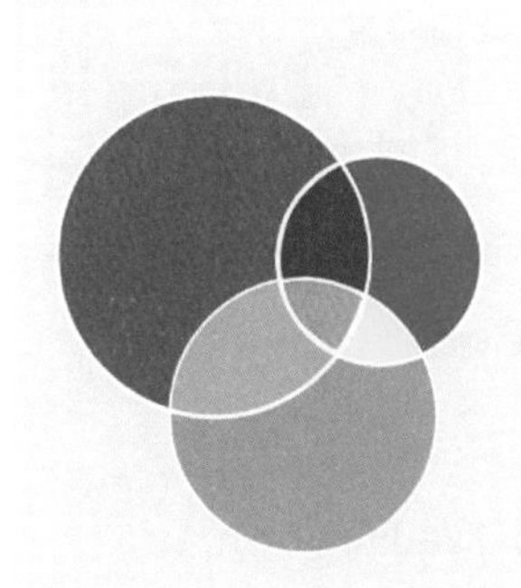

Play with Your Food

Ideas to investigate and games to play when you're eating, cooking, or party planning.

Contents

Play with Your Food in other sections

What's Inside?

How many seeds are inside a melon? 100? 1,000? Open one up and find out!

❶ Estimate how many seeds

Cut and set out the melon.

Are there more than 10 seeds in all? more than 100?

Talk About...

❷ Count

Remove, arrange, and count the seeds.

Levels: Easy, Medium

Group size: small enough so everyone gets a chance to predict and count

Materials:

- **Easy**. melon slice
- **Medium**. whole melon cut open so everyone can see the seeds

knife (adults only)

❸ Compare estimate and count

Were there more or fewer than you predicted?

Variations

Do all apples have the same number of seeds? (Easy). Gather a few apples of different size and type. Predict, cut, and count. Or, keep track of the number of seeds in each apple you eat over a month.

Edible explorations (Easy). Gather several types of fruit. Predict how many seeds in each, then cut and count. Do the larger fruits always have more seeds?

Pumpkin party (Medium). How many seeds in a Halloween pumpkin? Estimate, then cut and count before carving.

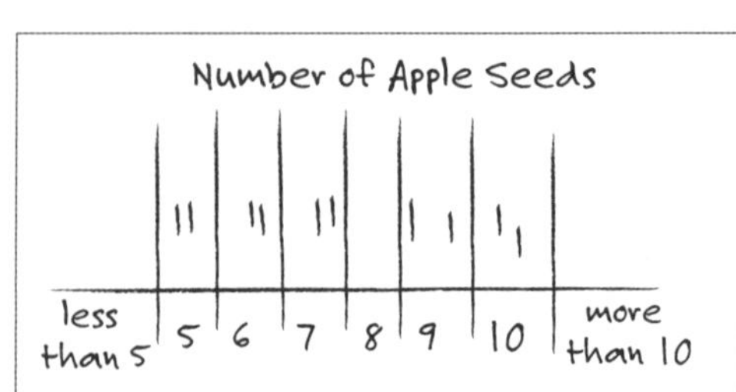

Can you pour out a "serving size" to match the one on the cereal box?

Level: Medium

Group size: small enough so everyone can pour out an individual bowl of cereal

Materials:

- cereal packaged with Nutrition Facts label listing "serving size" in cups
- measuring cup
- cereal bowl

1 Read the label

Find the serving size.

2 Say when

Fill a bowl until you think you've reached the serving size.

3 Measure

Use a measuring cup to find out if you were over, under, or just right.

4 Eat!

Pour the cereal back into the bowl and enjoy!

Variation

What's your serving size? (Medium).

Do you usually eat more than, less than, or about the same as the "serving size" listed on the cereal box?

Fill your cereal bowl with the amount you usually eat. Then, pour the cereal into a measuring cup to check.

Play with Your Food

Play a game like "Twenty Questions" using Nutrition Facts labels from your favorite foods.

Level: Hard

Group size: 3-5 per game; see Variation for a 2-player game

Materials:

about 20 Nutrition Facts labels cut from different food packages; include food names

Set up for the game

Lay out the labels face-up.

Decide who will be the Leader. The Leader secretly picks a label without removing it from the layout.

Take turns. On each turn:

1. **Ask a yes-or-no question to help figure out the secret food label. You may not ask if a certain label is the secret one.**
2. **The Leader answers the question and removes any labels that were ruled out.**

What's a yes-or-no question that could rule out half of the labels?

3. **The player who narrows down the labels to the secret one wins.**

Variation

Play with two people (Hard). Play two games. Each game, a different person is The Leader. The person who narrows down the labels to the secret one with the fewest questions wins.

Food Fight

Play a game like "War" using Nutrition Facts labels from your favorite foods.

Levels: Medium (Hard)

Players: 2-3 per game

Materials:

30-40 Nutrition Facts labels cut from different food packages; include food names

Set up for the game

Divide the labels into equal piles: one for each player.

Turn the piles face-down.

Take turns. On each turn:

1. **All players turn up the top label in their piles.**
2. The player whose label shows the highest protein per serving takes the cards played. In case of a tie, everyone turns over another card. The player with the highest takes all cards played.
3. Play until the piles are used up. The player with the most cards wins.

Which of the foods have a lot of protein? What's a reasonable amount of protein to eat each day?

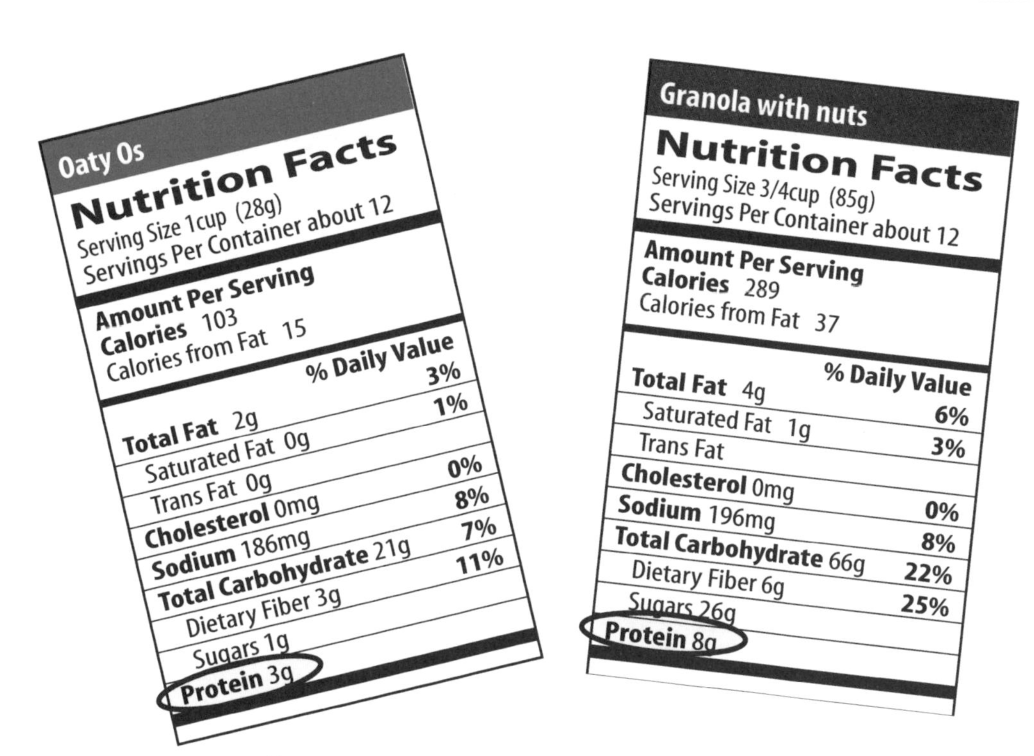

Variation

Fat fight (Hard). The player whose label shows the lowest percent fat per serving takes the cards played. Or, play for the lowest percent sodium or sugar.

Fair Shares

Snack time? Make sure everyone gets a fair share.

Levels: Easy, Medium

Group size: two or more sharing food

Materials:

countable food items, such as crackers or baby carrots

- **Easy**. 10-20
- **Medium**. 20-40

serving supplies (plates, napkins)

❶ Predict each share

If we share these, could everyone get two pieces? ten pieces? Why do you think so?

❷ Divide up the total

Deal them out, count them out, or divide and distribute.

❸ Compare predictions with results

Did you get more than you predicted? the same amount?

❹ Decide what to do with the extras (optional)

Cut them up and distribute fairly or save them for another time.

Variations

Unequal shares (Medium). Decide how to divide the food fairly if one person wants less.

One left over (Medium). Decide how to divide the food fairly so that there is one item left over in case a guest arrives.

Party Planning

Plan a party or special snack to fit your budget.

Levels: Medium (Hard)

Group size: small enough for everyone to have a say in the list

Materials:

- grocery store circulars or access to online grocery store price listing
- paper and pencil
- calculators

❶ How many and how much?

Find out how many people are coming and the total you can spend.

How much can you spend per person?
Is $1 each enough?

❷ What would you buy?

Look through grocery store circulars or go online to view price lists.

Make a list of what you'll buy.

How did you make your choices?

How did you stay within budget?

❸ Shop, cook, serve, and eat

Take your list to the store and shop. Enjoy the party!

Variations

Party favors (Hard). Figure in the cost everything you'll need for the party. Decide how much you can spend per person total on food, place settings, take-home bags, and other party favors.

Family dinner (Medium). Decide how much you will spend per person or how much in total for a family dinner. Then, plan the menu, and shop, cook, and eat!

Size Them Up

Which holds the most? Put containers in order by which would hold the most water.

Levels: Easy, Medium (Hard)

Group size: small enough so everyone has a turn to line up the containers

Materials:

5-8 containers (food, shampoo, or art supples) of different shapes and sizes, marked by capacity in fluid ounces

stick-on notes: use them to cover the capacity amounts (optional)

❶ Set out the containers

Mix the containers up, so they're not in size order.

❷ Predict

Which container do you think would hold the most water? Why?

Does the tallest always hold the most?

❸ Size them up

Put the containers in order from the one you think holds the most to the one you think holds the least.

❹ Check your order

Easy. Fill the container you predict holds the most. If it really holds the most, there will be some extra when you pour the water into the next largest.

Medium. Check the capacity amounts on the container labels.

Variations

Find the tallest (Easy). Line up the containers from tallest to shortest.

Estimate capacity (Hard). Find the smallest container. Read off the capacity in fluid ounces. Use that measure to estimate the capacity of the other containers.

Double or More

Try this when you're cooking for a crowd. Make two, three, or four times a recipe.

Levels: Easy, Medium, Hard

Group size: small enough so everyone has a chance to measure

Materials:

a recipe and related ingredients and supplies

- **Easy**. Choose a recipe to double. If you need to make enough for eight, use a recipe for four.
- **Medium**. Choose a recipe to triple or quadruple.
- **Hard**. Choose a recipe to increase several times. Use a recipe with several different fraction measurements.

❶ Who's coming?

Find out how many you need to feed and how many the recipe serves.

Smoothie for 2
1 banana
1 ¼ cups orange juice
½ cup frozen blueberries
5 frozen strawberries

How much will you need to increase the recipe to make enough for everyone? How do you know?

❷ How much of each ingredient?

Increase the amount of each ingredient by counting, measuring, adding, or multiplying.

How much will need to increase the recipe to make enough for everyone? How do you know?

Talk About...

❸ Make the recipe

Enjoy the results!

(continued on next page)

Variations

Count it out (Easy). Use a recipe that involves only whole-number amounts.

Half or less (Hard). Make 1/2, 1/3, or 1/4 the recipe.

Double the bubbles/dough (Easy, Medium, Hard). Use a recipe for bubble soap or play dough.

Bubble Soap for one

¼ cup liquid dishwashing detergent
¾ cup cold water
5 drops of glycerin

Good for Groups

Icebreakers, party games, challenges, and contests designed for a crowd.

Contents

Good for Groups in other sections

Animal Olympics

Crawl like a crab, creep like a cat, or run like a cheetah. How far in five seconds?

Levels: Medium (Easy)

Group size: 3 or more

Materials:

watch or clock that shows seconds

chalk or masking tape

ruler or tape measure

Before beginning

Mark a starting line with chalk or masking tape.

❶ Get in character

Choose an animal. Practice moving like that animal.

❷ Predict

Everyone lines up along the starting line.

How far do you think you can go in five seconds when you're moving like your animal? Make a prediction.

Talk About...

(Optional) Get a sense of five seconds by counting together: one Mississippi, two Mississippi, three Mississippi, four Mississippi, five Mississippi.

❸ Go!

Move like your animal while someone times five seconds.

❹ Compare or measure

Mark your ending point with chalk or tape.

Did you go as far as you thought in five seconds?

Talk About...

Closest prediction wins.

Variation

Cross the room (Easy). Could you cross the room in five seconds when you're moving like your animal? Could you get half way across the room? Predict, then try it.

Jump, Clap, Snap

Jump, clap, and snap on time or you're out. Last one standing wins!

Levels: Easy, Medium, Hard

Group size: 4 or more

Materials:
none

❶ Get in a circle and start counting by 1s

Easy. Jump whenever you say 10, 20, or another multiple of 10.

Medium. Start with the Easy version, and then add another action: clap whenever you say 2, 4, 6, or another even number. On multiples of 10, jump *and* clap.

Hard. Start with the Medium version, and then add another action: snap your fingers whenever you say 3, 6, 9, or other multiples of 3.

Which numbers get a jump, snap, and clap? Talk About...

If you miss a jump, snap, or clap, you're out.

❷ Keep going until just one person is left

Last one standing wins.

Variations

Everyone wins (Easy, Medium, Hard). Everyone stays in the game. Keep going until the group reaches 100.

Add another action (Medium, Hard). For instance, stamp your foot on multiples of 4. For extra challenge, count by 1/2s, 3/2s, 3s, 5s, or 7s instead of 1s.

Far and Wide

If we stood on each others' shoulders, could we reach the ceiling?

Levels: Easy, Medium, Hard

Group size: 6 or more

Materials:

ruler or measuring tape (Medium, Hard)

calculator (optional)

Before beginning

Decide on a question about how far, long, or high the group could reach together.

Easy. Choose something the group can try out:

If we lie head to foot, could we reach across the playground?

Medium. Choose something the group can figure out by trying out or measuring:

If we all joined hands and stretched out, could we reach all the way down the hallway?

Hard. Choose something the group will need to estimate:

If we stood on each other's shoulders, could we reach the top of the building?

❶ Predict

Could we reach that far? Why do you think so?

Talk About...

❷ Find out!

Try it or estimate.

Compare predictions and results.

Variation

Could we reach 1,000? (Hard). Make a prediction about 1,000. For instance, if we all stood on each other's heads, would we be more than 1,000 inches tall? If we all got on a giant scale, would we weigh more than 1,000 pounds?

Find Someone

Try this as an icebreaker, or play the game variation—fastest finder wins!

Levels: Easy, Medium, Hard

Group size: 4 or more

Materials:

- copy of "Find Someone" list for each person
- pencils
- rulers (Medium, Hard)

Before beginning

Make up a "Find Someone" list. Include about 10 items.

Easy. Include items about counting, comparing, and basic shapes (e.g., circle).

> Find someone
> who lives with more pets than people.
> wearing more than 5 buttons.

Medium. Include Easy items and some times about measurements.

> Find someone
> with hair about 6 inches long.
> more than 5 feet tall.

Hard. Include Medium items and some items about shapes and patterns on clothing.

> Find someone
> wearing parallel lines.
> wearing a shape with more than 4 sides.

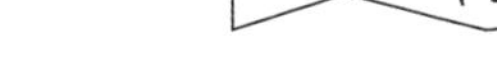

❶ Find someone

Everyone gets a list. Write down who you find for each item.

There could be some items no one fits.

❷ Share findings

Who did you find wearing parallel lines? Is anyone else wearing parallel lines?

Variation

Find it fastest (Easy, Medium, Hard). Play it as a game. First person to find someone for each item wins.

Quick Questions

Ask a quick question to break the ice.

Before beginning

Come up with a multiple choice question the group will enjoy answering. Write the question at the top of a large sheet of paper.

Put possible answers along the bottom of the paper.

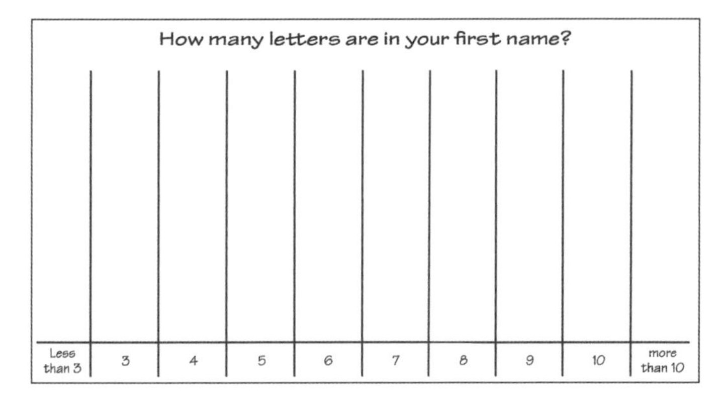

Levels: Medium (Easy, Hard)

Group size: 6 or more

Materials:

large piece of paper

markers or stickers

❶ Predict

What do you think the most common answer will be? the least common?

❷ Everyone answers

Use markers or stickers to show your answer.

❸ Compare predictions and results

Look over the answers. Any surprises?

What's your favorite season?

Fall | Winter | Spring | Summer

Variations

Yes or no (Easy). Use a question with just two answers. For instance, "Do you have a pet?"

Ask two groups (Medium). Each group responds in a different color. For instance, adults in red and children in blue. Do the groups answer in the same way?

It's about time (Hard). Use a question that involves time or another measurement. For instance, "How many hours of sleep did you get last night?" (6 or less, 6.5, 7, 7.5...) Or, "How tall are you?" (48 inches or less, 49, 50 ...).

Curtains Up

Curtains up! You have ten minutes to make up a skit using everything in the bag.

Levels: Medium (Hard)

Group size: 6 or more

Materials:

watch or clock

per group of 3-4:
one paper bag, two objects (e.g., penny, hat), a ruler, two index cards, and (optional) a world records book

Before beginning

Make up skit bags. Put in each bag: two objects, a ruler, two index cards, and optionally, a world records book. On one index card, write a measurement. On the other, write a number or price.

❶ Distribute bags and plan skits

Groups have ten minutes to plan a short skit using everything in their bags.

The skit must contain a disagreement about sizes.

❷ Perform

Set a maximum of five minutes per skit.

How would you resolve the disagreement?

Variations

Use the news (Hard). Include information you cut from a newspaper in each skit bag. Use three numbers, sports scores, or temperatures and three movie titles or headlines for each bag.

Tall tales (Medium, Hard). Each skit must include three claims involving sizes or numbers. Two must be possible (e.g., my brother weighed 7 pounds at birth), and one must be impossible (e.g., my sister weighed 20 pounds at birth). The audience needs to guess which is impossible. Try to trick the audience!

Two Way Split

Got a big group? Divide into two teams.

Levels: Easy (Medium)

Group size: 6 or more

Materials:
none

❶ Predict

Can we break into two equal groups? How do you know?

Talk About...

❷ Team up

Pick a way to divide into two groups (e.g., count 1, 2, 1, 2 and break into 1s and 2's; count off 1, 2, 3, 4 for a group of four).

Then, split!

If you can't make two equal groups, decide what to do.

Variations

Three way split (Medium). Predict whether the group could form three equal teams. Talk about different ways to find out and choose one. Try it and see.

Station rotation (Medium). Breaking into groups so you can rotate among activity stations? Come up with a rotation plan: how much time each group spends at each station, and the order in which they rotate.

Follow the clues to find five puzzle pieces. First to find and assemble them wins.

Levels: Easy, Medium

Group size: 1 or 2 per team

Materials:

- letter-size envelopes: 5
- clue list: 1 per team
- recipe, image, or special message for the puzzle: 1 sheet per team
- rulers
- tape
- recipe ingredients (optional)

Before beginning

Create a clue list of five clues. Each clue should lead to the place you will hide an envelope of puzzle pieces.

Easy. Use clues about sizes and counting.

1. Look under the box with 10 crayons.
2. Find the tallest chair. Look under it.

Medium. Use clues about measuring and shapes.

1. Look 10 inches above the book shelf.
2. Find the window with 9 rectangles. Look below it.

Make the puzzle pieces.

- Cut each puzzle sheet into five pieces.
- Put all copies of the same piece into an envelope. Write the number of the clue on the envelope.

Hide the envelopes in the locations that match the clues.

❶ Look for puzzle pieces

Each team gets a clue list. To avoid crowding at any one location, each team finds clues in a different order. When they find the envelope for a clue, they take one puzzle piece.

How did you count the window panes? Did you count squares as rectangles?

Each team finds five puzzle pieces.

❷ Assemble the pieces

Teams tape the pieces together. First to finish wins.

❸ Make the recipe (optional)

If the puzzle is a recipe, cook and eat!

Line Up

Pass the time when you're waiting in line—or play the Detective game (Variation).

Level: Easy

Group size: 4 or more

Materials:
none

Before beginning

Choose a size or number characteristic everyone can see and compare, such as hair length, height, or number of shirt buttons.

❶ Predict

If we line up from shortest to longest arm span, do you think you'll be at the start, middle, or end of the line?

❷ Line up

Make comparisons and stand in order. If two people have the same measurement, they stand side by side.

❸ Are we in order?

Check and change places if needed. The last person in line chooses how to line up next time.

Variations

Detective (Easy). On each round, one person, the Detective, leaves the room. The others line up in order by a secret characteristic they pick (e.g. number of buttons). The Detective returns and asks yes-or-no questions to figure out the characteristic (Is it about clothing? Is it about length?).

Play enough rounds so everyone gets a turn as the Detective. Keep track of how many questions each Detective asks. The Detective who asks the fewest questions wins.

Ready, set, go! You have two minutes to build a tower.

Levels: Easy, Medium, Hard

Group size: 1-2 per tower

Materials per tower:

- 20-30 containers of different sizes, small boxes, or other objects that can be stacked
- ruler, yardstick, or tape measure: 1 per tower (Medium, Hard)
- clock or watch that shows minutes (for whole group)

❶ Practice

Use the materials to practice building a tower that meets the goal.

Easy. Build the tallest tower you can.

Medium. Build a tower as close as possible to 4 feet high.

Hard. Build a tower at least 4 feet high with the smallest possible "footprint" (area in contact with the floor).

Which shapes make for a good tower base? Which work well in the middle?

❷ Get ready, get set

Take apart your practice towers.

Choose someone to be the timekeeper for the tournament.

❸ Go!

You have exactly two minutes to build a tower that meets the goal.

❹ Find the winner

The tower closest to the goal wins.

(continued on next page)

Variations

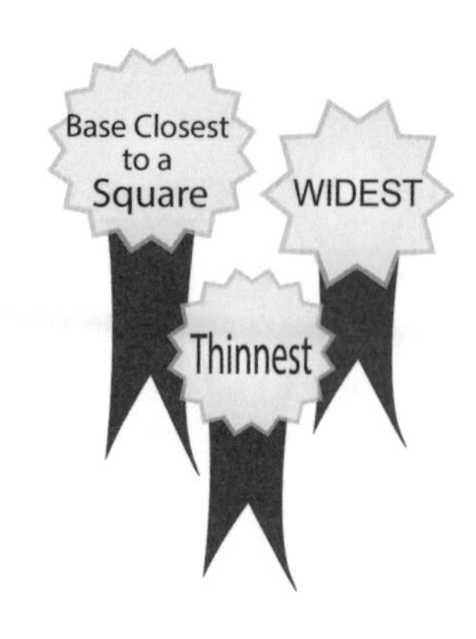

Everyone wins (Easy, Medium, Hard). Each tower wins an award for size or shape. Everyone decides on awards together.

Less is more (Medium). Build the tallest tower using the fewest items.

Which holds the most? (Hard). Make a silo. Build the tower that would hold the most grain.

Symmetrical designs (Hard). Build the tallest symmetrical tower.

Anytime, Anywhere

Activities to do and games to play wherever you are: in the car, on the bus, in a waiting room, or at the dinner table.

Contents

Anytime, Anywhere in other sections

Games Galore

Good for Groups

Take Stock

Waiting around? Pass the time by taking an inventory of legs in the room.

Levels: Easy, Medium, Hard

Group size: 2 or more

Materials:

none

❶ Estimate

Talk About...

How many legs are in the room: about 10? 100? 1,000?

Easy. Just count legs on people.

Medium. Count legs on people and other living things.

Hard. Count legs on people, other living things, toys, and furniture.

❷ Count

Find the number of legs in the room.

❸ How many did you find?

Compare answers.

How did you get your total? Did you count by 2s or 4s? add? multiply?

If answers differ, check by finding the total together. Make sure to account for anyone who might have left or arrived. The person who comes closest to the group answer wins.

If answers are the same, everyone wins.

Variation

Legs over time (Easy, Medium, Hard). If you'll be there a while, keep count every 15 minutes. When did you find the largest number of legs? the smallest?

What's on the Wall?

Need to fill a few minutes? Play a game on the wall!

Levels: Easy, Medium, Hard

Group size: 2 or more

Materials:

none

❶ Pick a wall and decide what to look for

Easy. Find as many circles (or squares) as you can on the wall.

Medium. Find as many triangles as you can on the wall.

Hard. Find as many rectangles as you can on the wall and the window.

❷ How many did you find?

Compare answers.

Does a square count as a rectangle? Why or why not?

If answers differ, check by finding shapes together. The person who comes closest to the group answer wins.

If answers are the same, everyone wins.

Variations

Without a wall (Easy, Medium, Hard). Try this with a billboard, magazine page, or scenic view.

Numbers on the wall (Easy, Medium, Hard). Find the largest number on the wall. What does the number show (e.g., a date, price, sports score)? For more challenge, find the smallest number on the wall. Include fractions, decimals, and negative numbers.

Rate It

How would you rate today on a scale of -2 to 2? Use ratings as a conversation starter.

Before beginning

Pick something to rate, such as the weather, your day so far, or a book everyone has read.

Levels: Easy, Medium, Hard

Group size: 4 or more

Materials:
none

❶ Decide on a rating scale

Easy. 1 to 5.

Medium. -2 to 2.

Hard. -5 to 5.

What would a rating of 0 mean?

Talk About...

❷ Rate it

Take turns giving ratings and explaining them.

❸ How did we rate?

What was the highest rating someone gave? the lowest? the most common?

Variation

Stories by the numbers (Medium, Hard). Write or tell a story that involves rating daily events. A visit from a friend might be a "2," a lost dollar "-2," and a game of jump rope "0."

Take Ten

Make cleaning up into a game.

Levels: Easy, Medium

Group size: 2 or more

Materials:

ordinary clutter

Easy. 50 or fewer items

Medium. 100 or more items

❶ Estimate

How many things do we need to clean up: about 50? 100? 1,000?

❷ Take ten

Everyone puts away ten items (fewer if there aren't enough).

❸ Are we done?

Continue taking ten (fewer if needed) until everything is picked up.

❹ How many in all?

Count by 10s to find the total. The person with the closest estimate wins.

Variations

Take five (Easy). Take just five at a time, with a total of no more than 20 items.

How long does it take? (Medium). Everyone estimates how long it will take to put everything away: 2 minutes? 5 minutes? 10 minutes? One person times while the others clean up. Closest estimate wins.

Countdown

3, 2, 1 ... it's time! Count down to a special event in days, minutes, or seconds.

Levels: Easy, Medium, Hard

Group size: any

Materials:

calendar

calculator (Hard)

❶ Locate the dates

Find today's date and the date of the special event on the calendar.

MAY				
S	M	T	W	T
				1
4	5	6	7	8
11	12	13	14	15
18	(19)	20	21	22

Today

JUNE				
T	W	T	F	S
3	4	5	6	7
10	11	12	13	14
17	18	19	(20)	21
24	25	26	27	28

Summer!

❷ How much longer?

Find how much longer in one or more of these ways.

Easy. Use the calendar to count the number of days to the special event.

Medium. Find how much time remains in months weeks, and days.

If it's May 19 and the special day is the first day of summer, June 20, summer begins in

Does today count?

Talk About...

Hard. Find how many hours, minutes, or seconds remaining.

❸ Continue the countdown

As the special day approaches, count down every day.

Variation

Count down to the full moon (Easy). Each night, take a look outside to see the moon waxing.

Minute Madness

Make time fly when you're tracking how much you can do in a minute.

Levels: Easy, Medium

Group size: 2 or more

Materials:

watch or clock that shows minutes and seconds

paper and pencil for each person

Before beginning

Chose an activity you can repeat for a minute, such as jumping jacks or drawing stars.

Easy. Pick something you can do 10-20 of in a minute.

Medium. Pick something you can do 50 or more of in a minute.

❶ Predict

How many stars do you think you can draw in a minute?

❷ Keep track

One person times for a minute while everyone does the activity and keeps count.

❸ How many?

Compare predictions with results.

How did you count and keep track?

(continued on next page)

Minute Madness (cont'd)

❹ Repeat

Talk About...

Will you do more, fewer, or about the same number next time?

Try it and see!

Jaya
Jumping jacks first try
Estimate 30
Actual 65

Jumping jacks second try
Estimate 70
Actual 72

Variation

Estimate a minute (Medium). One person, the Timekeeper, times a minute while the others estimate. The Estimators raise their hands when they think a minute is up. The Timekeeper notes whose hands go up before a minute, whose at a minute, and whose after a minute. Once all the hands are up, the Timekeeper tells everyone the results.

Any Year Calendars

Things to do on familiar holidays (like July 4), less common holidays (like Backward Day), and any day. Includes a bonus set of ideas for celebrating 100 days.

January

Get moving all month long

1

New Year's Day

Make three wishes for the new year.

3

In five seconds, how far can you hop?

4

Trivia Day

What is the highest human jump ever recorded? How high can you jump?

6

How many steps does it take you to walk across the room?

9

In five seconds, how far can you run?

10

Do 15 arm circles every hour.

11

How many times can you throw and catch a ball without dropping it?

15

In five seconds, how far can you crawl?

18

Turn right three times, then turn left four times.

Martin Luther King Jr. Day

Falls on the third Monday in January. Do a good deed for someone else.

20

National Penguin Awareness Day

In five seconds, how far can you waddle like a penguin?

23

Count to 23. Clap on even numbers, jump on multiples of 5.

25

Opposite Day

Stand face to face with a friend. Hop backwards in opposite directions for five seconds. How far apart are you?

Chinese New Year

Falls in January or February. Giant pandas are native to China. How many still live in the wild?

28

In five seconds, how far can you go when you're taking baby steps?

30

Make up a dance that takes exactly 30 seconds.

31

Backward Day

How far can you walk backwards in 10 seconds?

For more ways to get moving, see *Animal Olympics* (p. 38), *Jump, Clap, Snap* (p. 39), *Far and Wide* (p. 40), *Piece It Together* (p. 45), *Line Up* (p. 46), and *Minute Madness* (p. 55).

February

Create and design to your heart's content

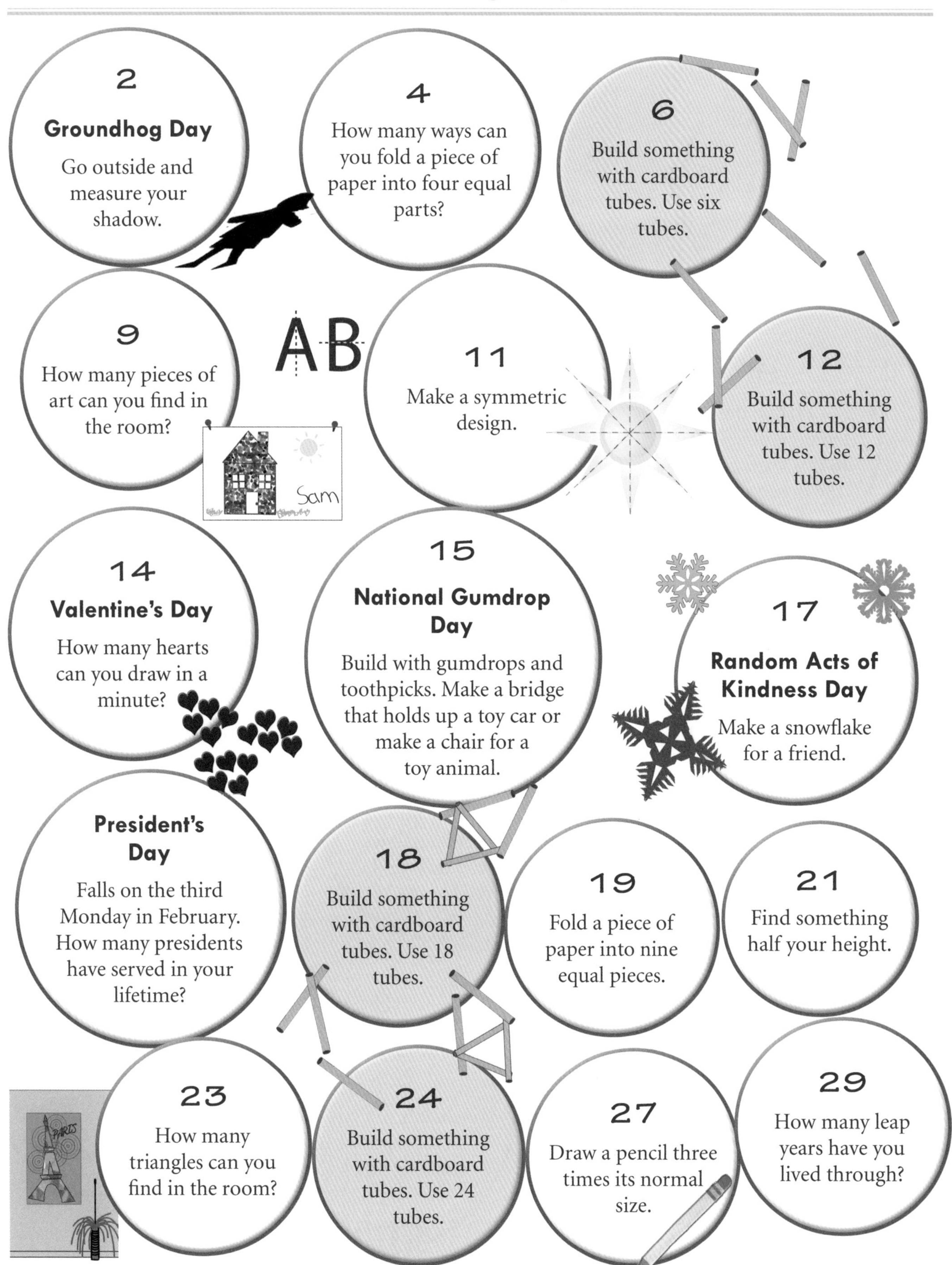

2
Groundhog Day
Go outside and measure your shadow.

4
How many ways can you fold a piece of paper into four equal parts?

6
Build something with cardboard tubes. Use six tubes.

9
How many pieces of art can you find in the room?

11
Make a symmetric design.

12
Build something with cardboard tubes. Use 12 tubes.

14
Valentine's Day
How many hearts can you draw in a minute?

15
National Gumdrop Day
Build with gumdrops and toothpicks. Make a bridge that holds up a toy car or make a chair for a toy animal.

17
Random Acts of Kindness Day
Make a snowflake for a friend.

President's Day
Falls on the third Monday in February. How many presidents have served in your lifetime?

18
Build something with cardboard tubes. Use 18 tubes.

19
Fold a piece of paper into nine equal pieces.

21
Find something half your height.

23
How many triangles can you find in the room?

24
Build something with cardboard tubes. Use 24 tubes.

27
Draw a pencil three times its normal size.

29
How many leap years have you lived through?

For more ways to create and design, see the **Projects and Crafts** section (pp. 15-26).

March

Watch the weather

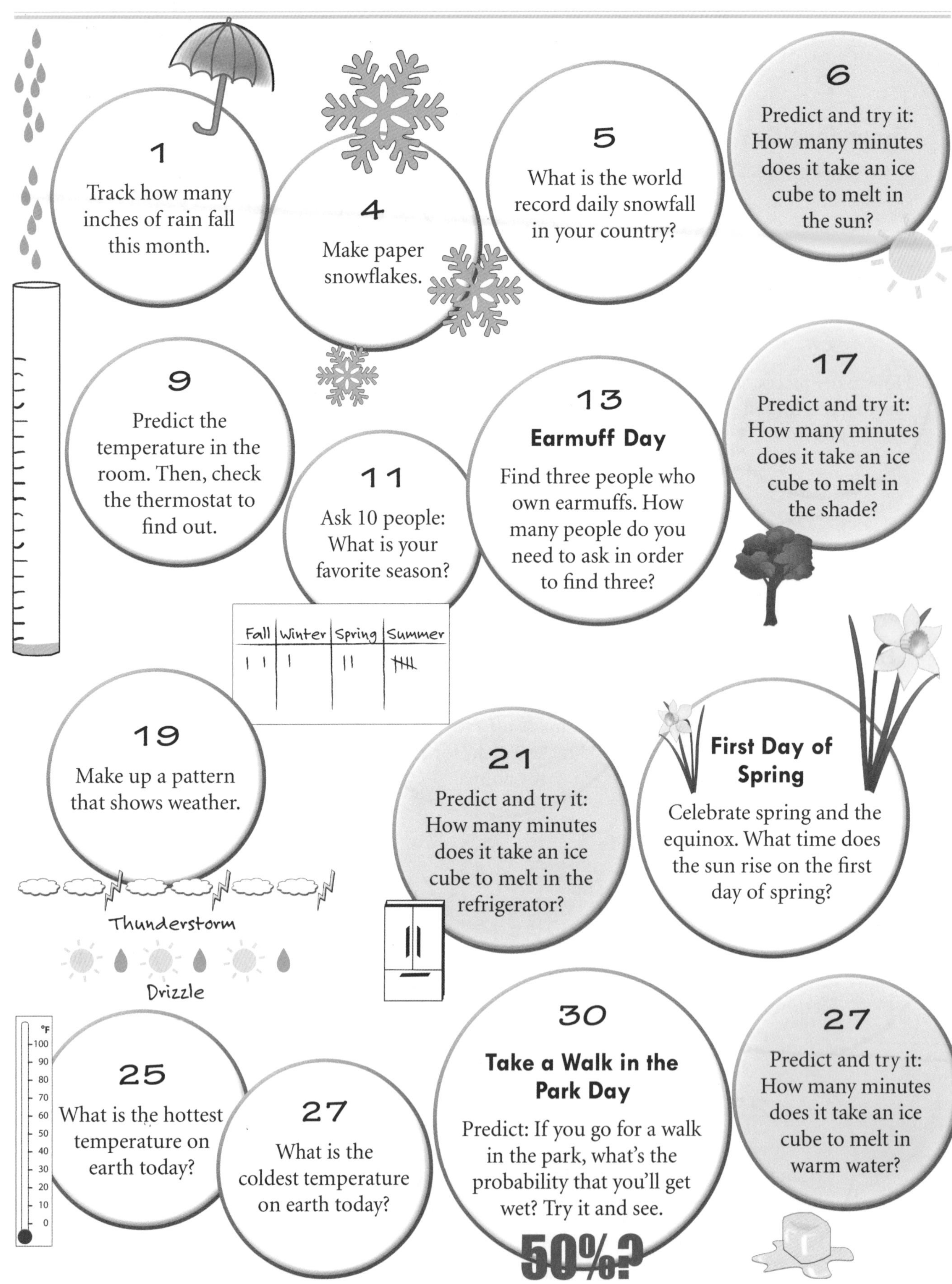

1
Track how many inches of rain fall this month.

4
Make paper snowflakes.

5
What is the world record daily snowfall in your country?

6
Predict and try it: How many minutes does it take an ice cube to melt in the sun?

9
Predict the temperature in the room. Then, check the thermostat to find out.

11
Ask 10 people: What is your favorite season?

13
Earmuff Day
Find three people who own earmuffs. How many people do you need to ask in order to find three?

17
Predict and try it: How many minutes does it take an ice cube to melt in the shade?

19
Make up a pattern that shows weather.

21
Predict and try it: How many minutes does it take an ice cube to melt in the refrigerator?

First Day of Spring
Celebrate spring and the equinox. What time does the sun rise on the first day of spring?

25
What is the hottest temperature on earth today?

27
What is the coldest temperature on earth today?

30
Take a Walk in the Park Day
Predict: If you go for a walk in the park, what's the probability that you'll get wet? Try it and see.

27
Predict and try it: How many minutes does it take an ice cube to melt in warm water?

Want more weather? Try *Rate It* (p. 52) to rate today's weather from -2 to 2. And don't forget to *Countdown* (p. 54) to spring.

April

Catch the beat with poetry and patterns

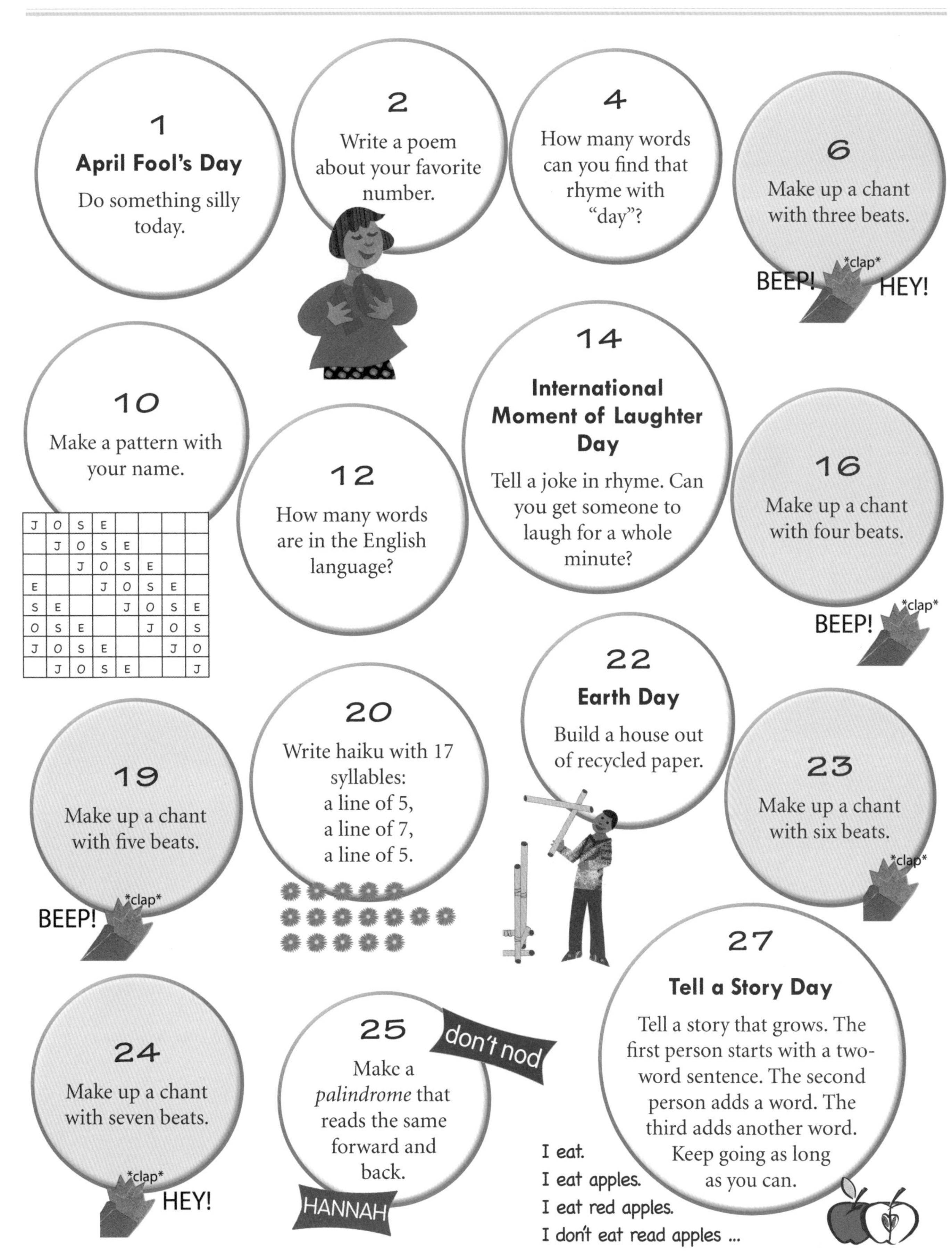

1
April Fool's Day
Do something silly today.

2
Write a poem about your favorite number.

4
How many words can you find that rhyme with "day"?

6
Make up a chant with three beats.

10
Make a pattern with your name.

J	O	S	E				
	J	O	S	E			
		J	O	S	E		
E			J	O	S	E	
S	E			J	O	S	E
O	S	E			J	O	S
J	O	S	E			J	O
	J	O	S	E			J

12
How many words are in the English language?

14
International Moment of Laughter Day
Tell a joke in rhyme. Can you get someone to laugh for a whole minute?

16
Make up a chant with four beats.

19
Make up a chant with five beats.

20
Write haiku with 17 syllables:
a line of 5,
a line of 7,
a line of 5.

22
Earth Day
Build a house out of recycled paper.

23
Make up a chant with six beats.

24
Make up a chant with seven beats.

25
Make a *palindrome* that reads the same forward and back.

27
Tell a Story Day
Tell a story that grows. The first person starts with a two-word sentence. The second person adds a word. The third adds another word. Keep going as long as you can.

I eat.
I eat apples.
I eat red apples.
I don't eat read apples ...

For more ways to explore patterns in words and numbers, see *Name Game* (p. 3), *Secret Number* (p 4), *Tell Me a Story* (p. 19), and *Jump, Clap, Snap* (p. 39).

May

Go back to nature

Need more nature? Play *Narrow It Down* (p. 13) with shells, leafs, or rocks; try *Growing Plants* (p 22); include facts about sizes of animals and plants in *Curtains Up* (p. 43); and save up for a wildlife charity with *Penny Jar* (p. 24).

June
Analyze animals

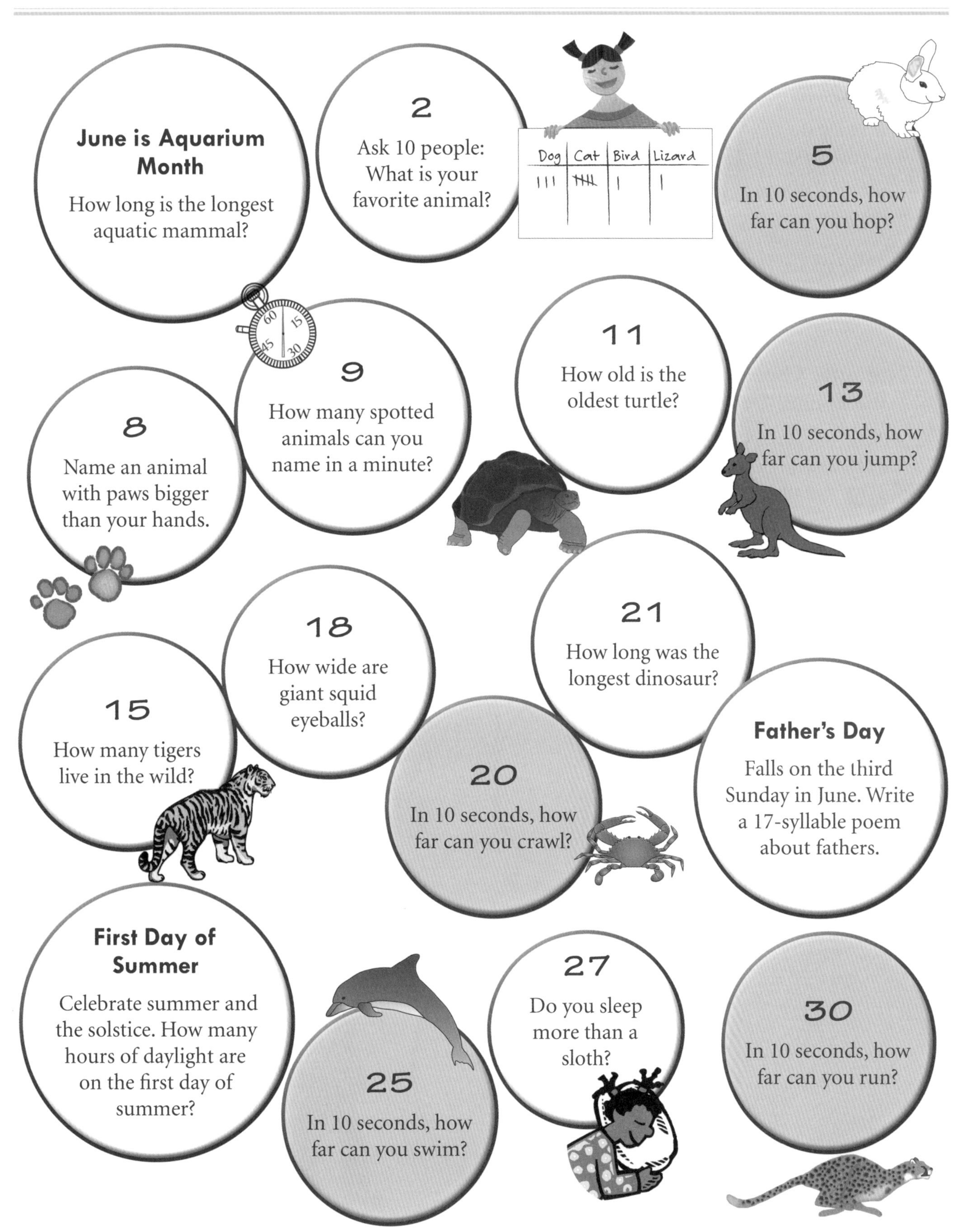

For more animal explorations, expand your world with *Giant Size* (p. 23) bugs and spiders, try *Double or More* (p. 35) with a wild bird food recipe, and get moving with *Animal Olympics* (p. 38).

July
Make cents

2
How many ways can you make 26 cents with different coins?

4
Independence Day
Going to watch a parade? Estimate how many people are in the crowd.

6
Collect coins for a charity every day this month. Donate at the end of the month.

7
How old is the oldest coin you can find?

11
National Cheer Up the Lonely Day
Play a coin game with someone who needs a friend.

12
Flip a coin 10 times. How many heads did you get?

13
How many times do you need to flip a coin to get 10 tails?

16
Keep collecting coins. Predict how much you'll have at the end of the month.

18
Flip three coins at once. How many flips until all three come up heads?

20
Find something in the room that costs between $1.00 and $5.00.

22
How many coins do you have so far?

23
Flip two coins at once. How many flips until both come up heads?

25
Count how much money have you collected so far.

26
All or Nothing Day
Flip a coin to make a decision today.

29
Make a final count. Does the total match your prediction? Donate!

For more to do with money, see: *Pennysaver* (p. 7), *Twenty Pennies* (p. 8), *More, Less, Equal* (p. 12), *Narrow It Down* (p. 13) with coins from all over the world, *Penny Jar* (p. 24), and *Party Planning* (p. 33).

August
Use your imagination

1
How many days are left until school starts?

3
What's the probability that you will go to the moon in your lifetime?

6
Build a tower as tall as you are. Who lives in it?

9
Book Lover's Day
Imagine you're a time traveler going back 100 years to meet someone your age. You can bring 10 books to share. What books would you bring?

11
Make up a story about the number 17 and an imaginary animal.

12
Ask 10 people: Would you rather be 20 feet tall or 20 inches tall?

14
Does the (very friendly) monster under your bed weigh more than you do?

15
What's the probability that someone with two heads will walk through the door in the next 10 minutes?

17
A giant beanstalk has sprung up on your usual route to school. Draw a map of a new route you can take.

19
What's the probability that you will be late on the first day of school?

20
If you were twice your height, could you reach the ceiling?

22
National Tooth Fairy Day
How many teeth have you lost?

23
Ride the Wind Day
Imagine you're a kite flying above your house. Draw a picture of what you would see when you look down.

26
What's the probability that the sun will rise tomorrow?

30
Use a piece of graph paper to design the floor plan of a treehouse.

Keep your imagination alive with *Build a Hideout* (p. 16), *Potato Bridge* (p. 17), *Ride on a Slide* (p. 18), *Tell Me a Story* (p. 19), *Giant Size* (p. 23), and *Curtains Up* (p. 43).

September
Travel the world

To weave the world into your day, play *Narrow It Down* (p. 13) with foreign coins or stamps, *Countdown* (p. 54) in a language other than English, and *Double or More* (p. 35) with a recipe from far away.

October
Crack the codes

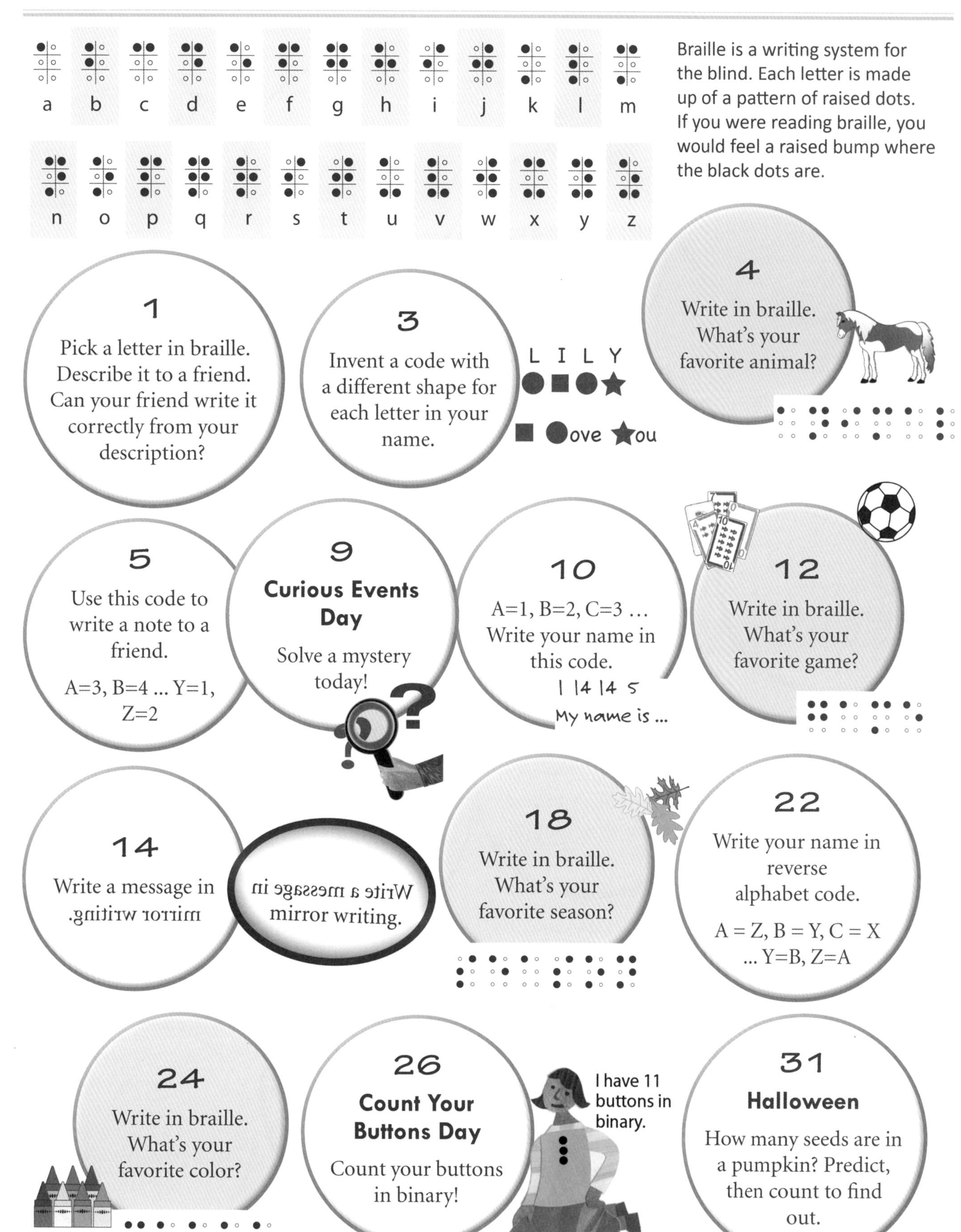

To continue with codes, write a counting book in code with *Tell Me a Story* (p. 19), and play *Name Game* (p. 3) using your name in code.

November

Master your menu

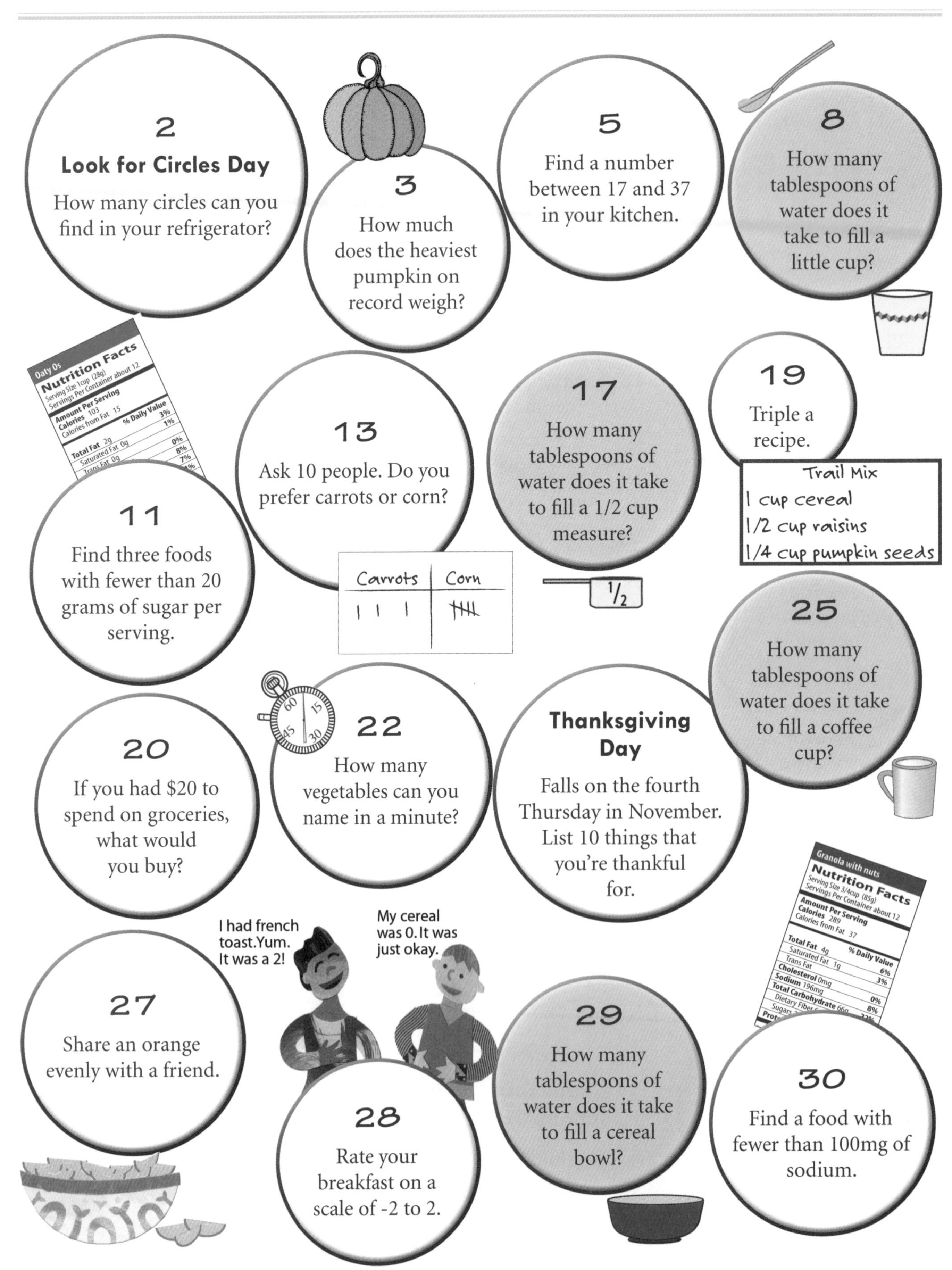

For more food activities see the **Play with Your Food** section (pp. 27-36).

December

Go for games

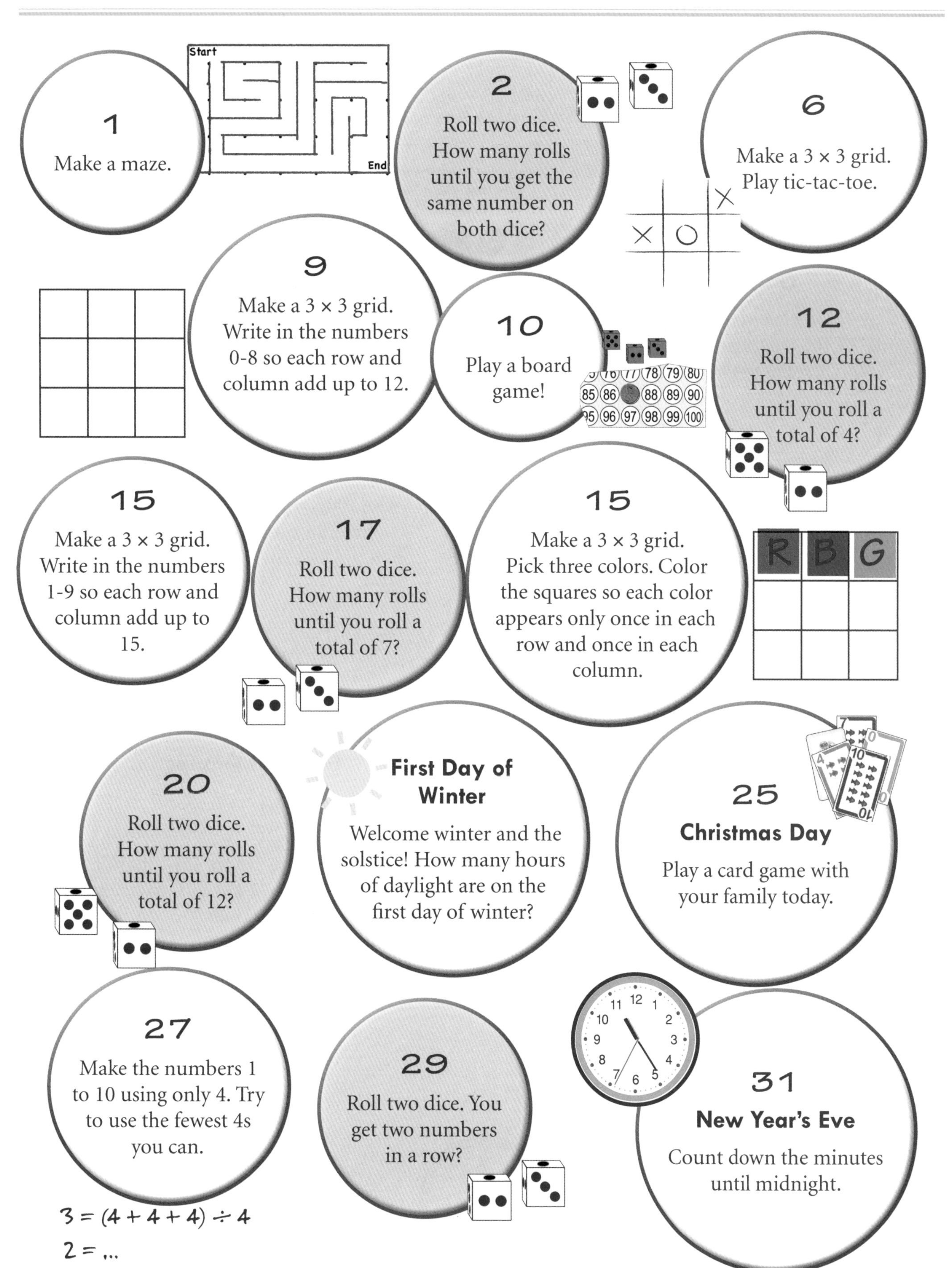

1
Make a maze.

2
Roll two dice. How many rolls until you get the same number on both dice?

6
Make a 3 × 3 grid. Play tic-tac-toe.

9
Make a 3 × 3 grid. Write in the numbers 0-8 so each row and column add up to 12.

10
Play a board game!

12
Roll two dice. How many rolls until you roll a total of 4?

15
Make a 3 × 3 grid. Write in the numbers 1-9 so each row and column add up to 15.

17
Roll two dice. How many rolls until you roll a total of 7?

15
Make a 3 × 3 grid. Pick three colors. Color the squares so each color appears only once in each row and once in each column.

20
Roll two dice. How many rolls until you roll a total of 12?

First Day of Winter
Welcome winter and the solstice! How many hours of daylight are on the first day of winter?

25
Christmas Day
Play a card game with your family today.

27
Make the numbers 1 to 10 using only 4. Try to use the fewest 4s you can.

$3 = (4 + 4 + 4) \div 4$
$2 = \ldots$

29
Roll two dice. You get two numbers in a row?

31
New Year's Eve
Count down the minutes until midnight.

Not done playing games? See the **Games Galore** section (pp. 1-14).

100 Days

Whether it's the 100th day of school, 100th day of the year, or 100 days since your birthday—here are some ways to mark the day.

Games

Play *Twenty Pennies* (p. 8) with a total of 100, *Land on 100* (p. 10), and *Secret Number* (p. 4).

Projects and Crafts

Tell Me a Story (p. 19) counting by 10s or dimes to 100.

Make it *Giant Size* (p. 23) by enlarging 10 times in two directions—100 times the area.

See how long it takes to collect $100 with *Penny Jar* (p. 24).

Use 100 toothpicks and 100 gumdrops to build a tower as tall as you can.

Play with Your Food

Which fruits have about 100 seeds? Predict and then cut, count, and eat with *What's Inside?* (p. 28).

Try *Party Planning* (p. 33) with a total of $100.

Good for Groups

Jump, Clap, Snap (p. 39) your way to 100.

Time for *Curtains Up* (p. 43) with skits about 100 minutes, 100 centimeters, and 100 pounds.

Piece It Together (p. 45) with clues about 100: find the number 100, something 100 centimeters high, and something that costs 100 cents.

Anytime, Anywhere

Countdown (p. 54) for the next 100 hours.

Have some *Minute Madness* (p. 55) in reverse: predict how long it will take you to draw 100 stars, then try it.

Math Standards: Common Core Connections

The games, projects, and activities in FOOD FIGHTS, PUZZLES, AND HIDEOUTS span the key topics in the Common Core State Standards for Mathematics for grades K-5. Use the charts on pp. 71-77 to match them, along with their variations, to the Mathematical Content Standards for each grade. Checkmarks on the chart reflect content children explore and strategies they typically use for each game, project, and activity in this book. Taken together, the calendars address the full range of Mathematical Content Standards for grades K-5.

Many children will benefit from experience with games, projects, and activities that address standards for earlier grades. The approach in this book can provide a new lens on familiar topics.

This book complements any math curriculum. You can use it to provide advance exposure to topics that children will be meeting in school, to give children hands-on experience with the skills they are currently learning in school, and to help them build enthusiasm about math and appreciation of its relevance to everyday life. All children—whether they are struggling or succeeding in school math—will benefit.

The resources in this book also offer a wealth of experience with the grade-independent Mathematical Practice Standards. For more information, see: National Governors Association Center for Best Practices, Council of Chief State School Officers. (2010). Common Core State Standards for Mathematics. Washington, DC: National Governors Association Center for Best Practices, Council of Chief State School Officers.

Common Core Connections (includes Variations)

		Kindergarten					Grade 1			
		Counting and Cardinality	Operations and Algebraic Thinking	Number and Operations in Base Ten	Measurement and Data	Geometry	Operations and Algebraic Thinking	Number and Operations in Base Ten	Measurement and Data	Geometry
Games Galore	Blockade, p. 2									
	Heads Up, p. 6	√	√				√			
	Land on 100, p. 10	√	√	√			√	√		
	More, Less, Equal, p. 12	√	√	√			√	√		
	Name Game, p. 3									
	Narrow It Down, p. 13				√	√			√	√
	Pennysaver, p. 7	√	√				√			
	Roll the Bank, p. 5									
	Same as Seven, p. 9									
	Secret Number, p. 4	√		√			√	√		
	Twenty Pennies, p. 8	√	√	√			√	√		
Projects and Crafts	Build a Hideout, p. 16				√	√			√	√
	Estimation Station, p. 20	√		√				√		
	Estimation Station Challenge, p. 21									
	Giant Size, p. 23									
	Growing Plants, p. 22				√				√	
	Penny Jar, p. 24	√	√	√			√	√		
	Potato Bridge, p. 17				√	√			√	
	Puzzle Me This, p. 25									
	Ride on a Slide, p. 18									
	Tell Me a Story, p. 19	√	√				√			

		Kindergarten					Grade 1			
		Counting and Cardinality	Operations and Algebraic Thinking	Number and Operations in Base Ten	Measurement and Data	Geometry	Operations and Algebraic Thinking	Number and Operations in Base Ten	Measurement and Data	Geometry
Play with Your Food	Double or More, p. 35	√	√		√		√			
	Fair Shares, p. 32	√	√				√			
	Food Fight, p. 31									
	Party Planning, p. 33									
	Play with Your Food, p. 30									
	Say When, p. 29									
	Size Them Up, p. 34				√	√			√	
	What's Inside? p. 28	√	√				√	√		
Good for Groups	Animal Olympics, p. 38				√				√	
	Curtains Up, p. 43									
	Far and Wide, p. 40	√	√		√		√		√	
	Find Someone, p. 41	√		√	√	√			√	√
	Jump, Clap, Snap, p. 39	√	√	√			√	√		
	Line Up, p. 46	√			√				√	
	Piece It Together, p. 45	√			√	√		√	√	√
	Quick Questions, p. 42	√			√				√	
	Tower Tournament, p. 47				√	√			√	√
	Two Way Split, p. 44	√	√				√	√		
Anytime, Anywhere	Countdown, p. 54	√	√				√			
	Minute Madness, p. 55	√	√		√			√	√	
	Rate It, p. 52				√				√	
	Take Stock, p. 50	√			√		√	√		
	Take Ten, p. 53	√	√	√			√	√		
	What's on the Wall? p. 51	√				√	√			√

Common Core Connections
(includes Variations)

		Grade 2				Grade 3				
		Operations and Algebraic Thinking	Number and Operations in Base Ten	Measurement and Data	Geometry	Operations and Algebraic Thinking	Number and Operations in Base Ten	Number and Operations — Fractions	Measurement and Data	Geometry
Games Galore	Blockade, p. 2	√			√	√	√			√
Games Galore	Heads Up, p. 6	√		√						
Games Galore	Land on 100, p. 10	√	√				√			
Games Galore	More, Less, Equal, p. 12	√	√	√		√	√			
Games Galore	Name Game, p. 3	√				√				
Games Galore	Narrow It Down, p. 13			√	√					√
Games Galore	Pennysaver, p. 7	√								
Games Galore	Roll the Bank, p. 5					√		√		
Games Galore	Same as Seven, p. 9	√				√				
Games Galore	Secret Number, p. 4	√				√				
Games Galore	Twenty Pennies, p. 8									
Projects and Crafts	Build a Hideout, p. 16			√	√				√	
Projects and Crafts	Estimation Station, p. 20	√							√	
Projects and Crafts	Estimation Station Challenge, p. 21									
Projects and Crafts	Giant Size, p. 23			√					√	
Projects and Crafts	Growing Plants, p. 22			√					√	
Projects and Crafts	Penny Jar, p. 24	√	√	√		√	√			
Projects and Crafts	Potato Bridge, p. 17			√						
Projects and Crafts	Puzzle Me This, p. 25	√		√	√	√			√	√
Projects and Crafts	Ride on a Slide, p. 18			√					√	
Projects and Crafts	Tell Me a Story, p. 19	√		√		√		√		

		Grade 2				Grade 3				
		Operations and Algebraic Thinking	Number and Operations in Base Ten	Measurement and Data	Geometry	Operations and Algebraic Thinking	Number and Operations in Base Ten	Number and Operations — Fractions	Measurement and Data	Geometry
Play with Your Food	Double or More, p. 35	√		√		√		√	√	
	Fair Shares, p. 32	√						√		
	Food Fight, p. 31								√	
	Party Planning, p. 33	√	√			√	√			
	Play with Your Food, p. 30									
	Say When, p. 29			√				√	√	
	Size Them Up, p. 34								√	
	What's Inside? p. 28	√				√				
Good for Groups	Animal Olympics, p. 38			√					√	
	Curtains Up, p. 43			√				√	√	
	Far and Wide, p. 40	√	√	√		√	√		√	
	Find Someone, p. 41			√				√		√
	Jump, Clap, Snap, p. 39	√	√			√		√		
	Line Up, p. 46									
	Piece It Together, p. 45			√	√			√		√
	Quick Questions, p. 42			√					√	
	Tower Tournament, p. 47			√	√				√	√
	Two Way Split, p. 44	√		√		√		√	√	
Anytime, Anywhere	Countdown, p. 54	√	√			√		√		
	Minute Madness, p. 55	√	√							
	Rate It, p. 52			√*			√*			
	Take Stock, p. 50	√		√						
	Take Ten, p. 53	√		√		√	√		√	
	What's on the Wall? p. 51	√			√					√

*Addresses negative numbers, included in "Number and Operations in Base Ten" Grade 6.

Common Core Connections (includes Variations)

		Grade 4					Grade 5				
		Operations and Algebraic Thinking	Number and Operations in Base Ten	Number and Operations — Fractions	Measurement and Data	Geometry	Operations and Algebraic Thinking	Number and Operations in Base Ten	Number and Operations — Fractions	Measurement and Data	Geometry
Games Galore	Blockade, p. 2					√					
	Heads Up, p. 6										
	Land on 100, p. 10										
	More, Less, Equal, p. 12	√									
	Name Game, p. 3	√					√				
	Narrow It Down, p. 13					√					√
	Pennysaver, p. 7										
	Roll the Bank, p. 5			√							
	Same as Seven, p. 9	√					√				
	Secret Number, p. 4	√					√				
	Twenty Pennies, p. 8										
Projects and Crafts	Build a Hideout, p. 16				√				√		
	Estimation Station, p. 20										
	Estimation Station Challenge, p. 21		√					√		√	
	Giant Size, p. 23			√	√				√	√	
	Growing Plants, p. 22										
	Penny Jar, p. 24	√	√		√			√			
	Potato Bridge, p. 17										
	Puzzle Me This, p. 25	√			√				√		
	Ride on a Slide, p. 18				√	√				√	√
	Tell Me a Story, p. 19	√		√					√		

		Grade 4					Grade 5				
		Operations and Algebraic Thinking	Number and Operations in Base Ten	Number and Operations — Fractions	Measurement and Data	Geometry	Operations and Algebraic Thinking	Number and Operations in Base Ten	Number and Operations — Fractions	Measurement and Data	Geometry
Play with Your Food	Double or More, p. 35	√		√					√	√	
	Fair Shares, p. 32										
	Food Fight, p. 31		√		√			√		√	
	Party Planning, p. 33	√			√						
	Play with Your Food, p. 30		√	√	√			√		√	
	Say When, p. 29										
	Size Them Up, p. 34				√					√	
	What's Inside? p. 28										
Good for Groups	Animal Olympics, p. 38										
	Curtains Up, p. 43			√	√				√	√	
	Far and Wide, p. 40	√	√		√		√	√		√	
	Find Someone, p. 41					√					√
	Jump, Clap, Snap, p. 39	√		√			√	√	√		
	Line Up, p. 46										
	Piece It Together, p. 45										
	Quick Questions, p. 42				√					√	
	Tower Tournament, p. 47				√	√				√	
	Two Way Split, p. 44										
Anytime, Anywhere	Countdown, p. 54	√									
	Minute Madness, p. 55										
	Rate It, p. 52		√ *					√ *			
	Take Stock, p. 50	√									
	Take Ten, p. 53										
	What's on the Wall? p. 51					√					√

*Addresses negative numbers, included in "Number and Operations in Base Ten" Grade 6.